Product Concepts

424 Inventive hand tool ideas
and other inventions for Students, Inventors,
Engineers and Creative people.

- ✓ If your struggling for products ideas this is the book for you.
- ✓ Gain an insight into product design.
- ✓ Gain a jumping off point to your own ideas. Find the creative genius in yourself.
- ✓ Patent your own ideas.
- ✓ Specific creative examples provided.
- ✓ Build creative thinking.

Dino von Noy

authorHOUSE®

AuthorHouse™
1663 Liberty Drive
Bloomington, IN 47403
www.authorhouse.com
Phone: 1-800-839-8640

First published by AuthorHouse 1/12/2010

ISBN: 978-1-4490-7138-7 (e)
ISBN: 978-1-4490-7137-0 (sc)

Library of Congress Control Number: 2010900044
Printed in the United States of America
Bloomington, Indiana

This book is printed on acid-free paper.

Important note about this book.

Thank you for buying my book.

I thought it worthwhile to share my product ideas and other inventions to help fellow creative people come up with there own ideas. In fact this book is to help fellow inventors and engineers springboard there ideas into useful products to help mankind in any way possible. In my life's research I have found it hard to get the insight of a creative person's methodology for creating ideas. Other people that might be interested in this book are Teachers, Visionaries, Architects and regular people who have some creative product idea in mind.

I believe we should share ideas in order to advance our humanity with group synergy. As example, if a reader of my book were to see one or more of my ideas then fuse that into something worthwhile that would be great. People of all ages could use this book to stimulate the creative side of the brain. My personal favorites are my ideas to help our military in foreign countries and the "Green" products to help the Earth.

I have intentionally left out the sketches and pictures of my ideas listed to help you think of what I'm talking about and possibly help you create your own abstraction of mine. This book is a benchmark whereby other creative people can build upon.

If you would like to buy my exact idea as written with further explanation and or sketches please write me at GREENS31@ HOTMAIL.COM. Please write "BOOK COMMENT" in subject line.

| **Idea number** | **Idea description** |

1. Help the U.S. Military with two products

I wanted to give something back to our country and so I propose two products to help our military in hostile countries.... My first product idea is to help stop improvised explosive devices (IED'S) on roads which our convoys travel. I call this product THE SMART ROAD...To embed NANO technology metallic granules into paving asphalt for roads. This new asphalt in combination with ground positioning satellites makes for a metallic grid vehicle road, readable from space...When an enemy digs a hole in this new road it's pinpointed and the satellite informs the military for counter action...Asphalt is a fast way to pave roads so the logistics of this idea is good.. My second product idea to prevent our soldiers from being killed by enemy snipers. This is a big problem in Iraq/Afghanistan right now...Our military gear for protecting the body is very good except for the head...This idea develops a new coating or device atop the helmet which gives off a glare (or flashing light) which prevents pinpoint viewing through the sniper scope...I believe this idea to be very feasible with today's technology.......

2. Use the "golden ratio" for hammer /tools design

This idea is to design a new generation of hammers and other tools based upon the "golden ratio" which says "a ratio within elements of a form, such as height to width approximating .618.... Example, the smaller element to the larger element $ab/ac = .618$....This golden ratio is found throughout nature in art and architecture...Examples of the golden ration are Seashells, the Stradivari violin, Leonardo da Vinci's paintings, The Parthenon, The great pyramid of Giza, Stonehenge and man's body (limbs).... When I benchmarked my collection of hammers I got the following readings 5-7/8" width of head to claw x 16" length of handle = .367 this is half of .618 so if we increase the head to claw to around 10 inches then this hammer should perform better in the human hand...Or decrease the hammer handle on this sample to get the .618 golden ratio....This is just one example of many that can be done with tools....My data is based on scientific face and observing that a lot of tools are out of scale......

3. The Fireproof towel

This product idea is an over sized bathroom towel made out of this new "oven glove" type material which can protect the typical consumer from severe heat...The uses of course would be for consumers to have these Fireproof Towels anywhere they live and work...In case of a fire the typical consumer throws this fireproof towel over there person giving themselves a few more seconds to walk by or through a flame to safety, thereby minimizing the severity of the flame...If one can somehow make this product relatively cheap then a typical family can have a fireproof towel for each member of the family...Another use might be to keep these new towels in the car to help anybody in a fire burn distress...I believe this fireproof towel is a noble product.....Fireman can also use this product when they enter burning dwellings to help the people stuck inside...Beyond this is fireproof bedspreads etc.

4. Soda can tab cools contents when opened

This idea ergonomically designs a soda can tab which triggers a cooling reaction in a chemical which is housed beneath the contents of the can...When the consumer buys a warm soda can they simply pull the tab to drink it and within a minute the content liquid is cold. The cold feature might last for 10 minutes or so, enough time for the consumer to swallow there drink cold...Great for outdoors and camping....Eliminates the need for continual refrigeration of soda cans. The actual chemical combination to make the liquid cold can might be like the cool ice products sold for injuries (ice pack ready when you open the packet, lasting a few minutes).....

5. Perfume and Shampoo bottles contents warmer

This idea ergonomically designs a small cheap mechanism which warms the contents of a shampoo or perfume bottle in order to give the consumer a nice warm feel from there product instead of that cold uncomfortable feeling when the product touches the human body.... This mechanism could be solar powered by artificial light too...It's not meant to last a long time and very small to fit inside the cap area of the bottle.....

6. Firestorm web netting w/hollow water channels

This idea is to design interlocking chain link web nets which can be as small or as large as desired.....This special linked netting is made of hollow core tubing in which water can run through and be sprinkled onto a dry ground condition via tiny holes throughout this special netting. The material for this net is green (eco friendly)...Now huge parcels of land can be covered with this product in order to prevent firestorms near cities and highways, such as what we are experiencing in Los Angeles and Orange counties of California right now...Prevention is the way to go on these mass disasters......I can visualize these nets being laid out close to our important highways which are near wild fire zones and near perimeter around areas of extreme fire hazards such as the Malibu mountains of California.....The experts predict our firestorms in the United States will get bigger as the planet changes and so it's a great idea to start this project now......

7. Home firefighting floating pool pump - Mobile

This idea benchmarks to the home firefighting pool pump as described below except it offers the ability for the mechanical device to float in the pool being pumped...It also has wheels for when the pool is empty of water it can be taken out and set into another pool....Right now as I write this we are being burned by firestorms in Los Angeles and Orange counties...Why not offer the high end home owner a tool which can float in there pool and be ready to water down any structure at a seconds notice?.... Respectfully....Benchmark company and details of there product as follows:...."" Our Home Firefighting® Cart Systems come fully equipped with everything you need to Defend Your Defensible Space® before the fire department arrives. Built around our high-performance Home Firefighting® Fire Pumps these carts deliver over 150 PSI (providing over 350' of vertical lift) and feature professional forestry fire hose, ruggedized aluminum professional firefighting nozzles and a professional aluminum

D-handle shutoff with rubber pistol grip for better control of the fire hose. Perfect for anyone with a pool, water tank, lake, pond or river. It's like owning your own fire truck! Maneuvers easily up and down rough terrain with minimal effort thanks to our exclusive balanced cart system that allows for moving the system with only one hand ! All Systems Feature: Home Firefighting® Fire Pump Briggs & Stratton Vanguard OHV Engine (Honda GX Engine Optional) Davey Firefighter® Twin Impeller Wet End 1.5" Heavy Duty Steel Roll Cage 1.5" Steel Handle w/Suction Hose Rack Double Powder Coated Finish Ballistic Hose & Equipment Storage Bag 12.5" Pneumatic All-Terrain Tires 1.5" Male Camlock Suction Fitting 3-Way Fire Hose Discharge Hard Coated Aluminum Caps 100' of Forestry Fire Hose (Click here for hose specs) Aluminum Pistol Grip Valve w/D-Handle Shutoff Aluminum Fog/ Stream Nozzle 16' Flexible PVC Suction Hose Brass Foot Valve & strainer...........

8. Utility knife blade with pull out capability

This idea ergonomically designs a mechanism which let's the user use a Utility knife as usual except with the flip of a lever this blade is easily detachable. This design is for more precise blade work...The blades on this knife are also hand fitted through the front because each blade comes with a wax sealer over the cutting edge...Now users don't hassle over undoing a Utility knife handle to get at a fresh blade....

9. Utility knife w/ flip out blade & stub

This idea designs a new generation of utility knife...This new knife is cheap except the blade flips opens like the new pocket knives except keeping the traditional handle....This new Utility knife also benchmarks to a skeletonized handle. It has a thumb stud for easy opening blade plus a pocket clip....I note the conventional utility knives have there place but the actual blade is too short for other works, why not combine the handle with the long blade of a pocket knife?....Please see the SCHRADE pocket knives in the internet, the ones with the thumb stub and skeleton body feature as benchmark.....

10. The 3 D Halogen projected tape measure reel

This idea ergonomically designs a triangulation feature on the typical tape measure which projects a halogen tape measure reel to a certain length, say 13 inches...This idea would be an awesome new concept in measurement...Instead of a tape measure coming out of the tape measure it would be a halogen projection of a tape measure reel...The limitation would be how much extension, I'm thinking only 13 inches or so...The novelty of this concept would be a selling point in itself....It would be nice to project a measurement in 3d space and onto objects....

11. The kitchen cabinet toolbox

This idea ergonomically designs a toolbox specifically to fit in kitchen cabinets...The exact dimensions to be determined but it will be more square in design and rather shallow...The user would simply flip open the door to there kitchen cabinet and there is there toolbox...This toolbox is mean to fit on top and bottom cabinets....This new toolbox will open through the front wall facing the user instead of the typical flip lid design....This new toolbox will also have a light "on" as soon as the front door is opened on it.....

12. Gravity fed wood nail dispenser pouch worn on belt

This idea ergonomically designs a synthetic pouch in which the user fills it up with any given wood nail or otherwise...The design of this pouch is worn on the typical pants belt and actually rides upside down on the belt...This pouch is designed so the nails feed down with only one nail at the pick out point by the human hand...Now users can fill up there pouch with nails orientated horizontally and have those nails ready for action anytime and anywhere....

13. Courtesy light at Utility knife blade

This idea ergonomically adds a light atop the area where the blade comes out of your utility knives. This courtesy light comes on as soon as the blade out button is pushed or pressed...he light could benchmark to those super long lasting flashlight blades on the market...This is a BRIGHT idea for the Utility knife of the future.....

14. The double helix level

This idea ergonomically designs a double helix level with two tubes like the one pictured...This idea is a level similar to your large levels like 48" except the ergonomic design is two inter-weaving tubes...The levels themselves are placed horizontally and vertically along these two tubes to produce a neat looking and functional level...The two interweaving tubes allows the user to place there hands anywhere along the tube geometry....

15. New pentagon blade design for new utility knife

This idea ergonomically designs a new blade for a new utility knife...This blade has five sides that are useable...The knife is designed whereby the pentagon blade is simply shifted to the next blade for sharp use.....Let's rock the utility knife world with this design...Now the user has more life out of the typical blade, five times more in fact.....

16. Combine generic hand tools & Survival kit for auto

This idea ergonomically designs an official tools kit to be carried in the trunk of the typical vehicle driven in the United States...This basic tool kit also includes an earthquake/fire/haz-ard survival kit...Now the drivers not only have a basic tool kit but also a survival kit ......The combined kits are housed ergo-nomically in a sturdy small suitcase type carrier...The survival kit could be for 3 days....

17. Colored street marker synchronized w/street signal

This idea ergonomically enhances pavement markers simi-lar to the ones used on highways and freeways to keep mo-tor vehicles in there lanes, except they have integrated color changing lights inside of red, yellow and green......These new lighted pavement markers are synchronized with the colors of the typical street signal overhead (red, yellow, green)...In other words when the intersection signal turns red, yellow or green so do these pavement markers which help the pedestrian cross an intersection more clearly and help the drivers themselves get a more visual picture of the signal mark....I note people in recent years kind of ignoring signals too...Now the public has two rows of street markers which light up to the color of the overhead street crossing......This is great for areas around schools....The marker itself could be designed like the signal above with 3 colored light lenses, each with it's respective color...Respectfully....Reference data on street markers:..."Question,,What are those little bumps in between the lanes on the freeway?...Answer:Those are known as raised pavement markers, or "Botts' Dots." In 1953, Dr. Elbert D. Botts, working in the Caltrans materials testing lab in Sacramento, came upon the idea of using a raised pavement marker to help make the painted lines separating lanes last longer. After a many refinements, the use of Botts' Dots were mandated for all California freeways, except in areas where they would be damaged in snow-removal operations. The ubiquitous little buttons have since been adopted around the world. In addition to making lanes easier to distinguish, the markers also had an additional -- and originally unintended -- safety benefit: to alert motorists when they drift out of their lane. There are an estimated 20 million Botts" Dots in place today on California freeways and highways -- a lasting legacy to Mr. Botts, who passed away in 1962.........

18. Hybrid power tools benchmarks to watch eco drive

This idea benchmarks to watches and the new eco drive solar power battery systems...This idea would have electrical tools with hybrid power back ups via this eco drive system...Now when the user does not have electrical power source they could possibly turn to there eco drive for a few minutes of work...... Reference data on eco drive solar power batteries:....Most Eco-Drive watches are equipped with a special titanium lithium-ion secondary battery that is charged by an amorphous silicon solar cell located behind the dial.[2] The titanium lithium-ion composition precludes it from being affected from the regular charge/

discharge cycle of other types of rechargeable batteries. Light passes through the covering crystal and dial before it reaches the solar cell.[3] Depending on the electronic movement model, a fully-charged secondary power cell could run with no further charging anywhere from thirty days to five years. If kept in the dark for too long, some movement models engage a hibernate mode, where the hands of the watch would stop running but the internal quartz movement would still keep track of time. If an ample supply of light is given, the hands would move to the proper positions and resume regular timekeeping............

19. Baseball bat design for new hammer handle

This idea ergonomically copies the design of a typical baseball bat and sizes it down in scale to be a handle for a new hammer...I believe the dynamics of a baseball bat will fair well as a hammer too....Only experiments will tell for sure...Respectfully.... Reference data on a baseball bat: Although using a stick to hit a ball is a somewhat simple concept, the bat is a complex object. It is carved or constructed very carefully to allow for a quick balanced swing, while providing power. The bat is divided into several regions. The barrel is the thick part of the bat, where the bat is meant to hit the ball. The part of the barrel best for hitting the ball with, according to construction and swinging style, is often called the sweet spot. The end of the barrel is not part of the sweet spot, and is simply called the tip or end of the bat. The barrel narrows, and becomes the handle. The handle is very thin, so that batters can comfortably set the bat in their fingers. Sometimes, especially on metal bats, the handle is wrapped with a rubber or cloth grip. Finally, next to the handle is the knob of the bat, a wider piece that keeps the bat from sliding out of a batter's hands. Over the centuries, the baseball bat's form has become more refined. During the 19th century, many shapes were experimented with, as well as handle designs. Today, the baseball bat is much more uniform in design. "Lumber" is a sometimes-used slang term for a bat, especially when wielded by a particularly good batter. The bat drop of a baseball bat is the difference of its weight (in ounces) to its length (in inches). For example; a 30-ounce, 33 inch long bat has a bat drop of minus

3 (30 - 33 = -3). Larger bat drops help to increase swing speed. Bats with smaller drops...........

20. Bow saw with looping blade (motorized)

This idea motorizes a bow saw so the blade can run (loop) in a clockwise direction and at variable speed to the users choice..... Now the user has the neat abilities of a fine cutting intricate geometry bow saw without having to actually sweat the hand sawing action.....The blade on this saw is flexible and able to move at the square design of this saw via a motorized mechanism..

21. Crescent claw hammer

This idea ergonomically designs a crescent wrench mechanism where the conventional claw is located of a typical hammer....In other words this idea is a claw hammer with a crescent wrench in lieu of the claw...Now the users have a tool upon they can grasp objects including nails as they tighten and pull....

22. Build a tool benchmarks to build a bear

This idea ergonomically designs a basic handle upon which various hand tools can be attached to configure different tools such as screwdriver, hammer, crescent wrench etc...Another electronic handle could be developed for power tools...The idea is the user can buy a generic handle and have the ability to buy specific use tool attachment....This idea benchmarks to BUILD-A-BEAR which is a children's toy whereby the kids can configure there own bear.....Reference data below.....Respectfully....:Build-A-Bear Workshop was founded by Maxine Clark in the year 1997. The first store opened in the Saint Louis Galleria. Ten years on, over 300 stores have been opened in countries worldwide: The United States (including Puerto Rico), Canada, Japan, Korea, Republic of Ireland, Denmark, Singapore, Germany, Sweden, The Netherlands, Australia, France, Russia, Taiwan, Thailand, Norway, and the United Kingdom. To celebrate the release of the 2006 film Happy Feet, plush toys featuring Mumble the penguin, the main character of the film, were created and sold. The toys were able to be fitted with a sound box that said four lines from the film.

On March 7, 2006, Build-A-Bear Workshop announced that it had given $1 million (US) to the World Wildlife Fund (WWF) through the sales of its WWF Collectibear stuffed animal series. For each plush animal sold, one dollar goes to the WWF to protect and conserve wildlife around the world.[1] On May 10, 2006, Build-A-Bear Workshop announced a line of toys based on its store would be included in McDonald's Happy Meals. It became the first plush toy collection included with the Happy Meal to feature removable mix-and-match clothing. Sixteen bears were created. They were released during the four weeks between May 12 and June 8. Each of the 16 bears were released in one of two outfits, unique to each bear, making 32 toys in all.[2] McDonald's ran a second Build-A-Bear Workshop Happy Meal in August, 2007.........

23. Electrical cord prong that locks into wall socket

This idea ergonomically designs an electrical cord on power tools which has a locking prong that engages the wall socket and cannot be pulled out without the user flipping a tang to open the cord end...The prong on this idea is slightly wider than the wall prong yet moves inward as it enters the wall socket...Only the user can unlock this cord prong...Now users don't have to worry about electrical cords being disengaged by accident while working and they move beyond the cord length...a sort of safety device...

24. Quarter height handle hammer

This idea designs a new hammer with a very short handle... Some craftsman prefer a short handle....As long as the golden ratio is followed this hammer should work just fine....This style could run on the full line of hammer types...This new hammer to also include the anti vibe handle.....Reference data: Item 79001: Great Neck Mini Hammer. Small and convenient with many uses. High carbon steel construction for strength and durability. Comfortable rubber grip ergonomically shaped handle. Weighs 14.5 oz. .....

25. Folding solar panel /hybrid power tools of future

This idea is an oversized folding solar panel for power tools of the future which would be hybrid...Several of these tools cold be hooked up to this panel when it is unfolded at the jobsite... Reference data:.. The Solaris series from Brunton is their most compact solar panel solution. Charge your cell phone, PDA, digital camera, or MP3 Player in only 2 to 4 hours. Just plug your appliance into the provided female cigarette lighter adapter, just like you would in your car. The multi-section folding panel design allows maximum portability to the flexible high performance CIGS solar cells. Each Solaris features reverse flow protection, and up to 3 panels can be linked together for even more power output. For example: while one 6W unit has a maximum output of 15.4 Volts at 430mA, two units together can offer a maximum output of 12 Watts: 15.4 Volts at 860mA. Each Solaris Solar Panel Array comes complete with a storage sack, female vehicle outlet cable, multi-linking cable, and a battery clamp cable. Overall Flat Dimensions: 29in x 9in (less than 1/32in thick) Folded Dimensions: 5in x 9in (approx 1/2in thick) Weight: 7.1oz Max Output: 6 Watts (15.4 Volts / 430mA) Mfr Warranty: 1 Year Panel Type: Multi-junction CIGS..........

26. Field compass w/ integral horizontal liquid level

One of the biggest problems with inexpensive compasses is getting the magnetic needle stuck on the edges and not correctly pointing north, one has to balance it horizontally....Everybody fidgets a lot with these compasses and so I thought why not integrate a horizontal liquid level into a typical inexpensive compass?....Now the user adjusts the liquid level so the compass balances out correctly and allows the magnetic north needle to freely point north....

27. Camping tool w/hollow handle that serves as stove

This idea adds an air tight hollow handle to some camp tool such as a folding shovel, axe or hammer...The hollow handle has small holes along it's length...The holes act as a flame spreaders like a cooking stove....The user places this tool down and fills the hollow handle with alcohol which they carry in another storage container....When ignited the holes on the handle act like a stove to heat food and other things. This idea benchmarks to an actual item which is described below...Now user don't have to buy expensive and bulky camping cooking stoves because it's already integrated into another tool.. .....Reference data:..These stoves are made from aluminum soda cans, and weigh about an ounce. I had read about these little gems here and there, and had planned on making one, but all the instructions out there seemed to involve taping the seams and other rather flimsy approaches. I don't like flimsy gear that could fail me at the wrong times! A guy called "Mechanic Mike" has solved this. He has made dies and tooling for use in a press to effectively produce these with no tape or adhesives, These are permanently pressed together, and are quite durable when compared to anything else out there. He sells these in his Ebay store (see link below) for about $5/ea. http://stores.ebay.com/Thru-Hikers-Closet-and-More_W0QQssPageNameZl4QQtZkm To use the stove, you just put in a small amount of alcohol (a few ounces, or enough to cover the base). I use a small mil-spec plastic flask to carry the alcohol. Light the alcohol, which will burn with a nice blue flame. As the stove walls heat up, they begin to vaporize the alcohol internally, and within about one minute, each of the 30 or so holes at the top is sporting a nice blue flame jet. The effect is like cooking on a gas stove at home. A good size canteen cup of water will boil in minutes. As you may have noticed, I am fond of compact, lightweight gear (SnugPak bags, bivy style tents, mult-tools etc). This piece of gear is about the nicest and cheapest piece of kit I have seen in quite some time. ...

28. Drooping handle pliers

This idea ergonomically adds a new drooping handle to the typical pliers tool.......I believe a stronger grip is achieved with this new droop design......Reference data on typical design to be modified with new handle..84-024 - 10" Bi-Material Groove Joint Pliers Print Printer Friendly Version Email to a friend Displaying product images/Previous Page Icon. gif 4 of 32 images/Next Page Icon .gif Click to Enlarge Click to Enlarge Features and Benefits *

Adjustable-width jaw designed for grasping and turning * Used for plumbing, automotive and general applications * Rust-resistant chrome finish * Machined jaws help grip items securely.....

29. Skeletal framed hand tools

This idea ergonomically designs hand tools with a very narrow interconnected skeletal frame with a lightweight yet rugged cover lining....For example, a crescent wrench and screwdriver could have this feature....This idea comes of course from the human body connections....I'm thinking cost would drop and so would weight in making these tools...

30. Taper handle on hand tools

This idea designs taper handles on hand tools. the taper is in reverse to what one thinks...Hammers, screwdrivers and wrenches can be designed this way...Exact taper angle to be determined.....I believe a more ergonomic feel would be achieved and lighter weights to tools.......Respectfully...Reference data Product Features: # World famous Crescent ® brand # For professional, industrial or home use # Tension spring stabilizes jaw and knurl adjusts easily # Alloy steel and heat treated # Proportioned dimensions for greatest strength and minimum weight # Chrome plated finish resists rust and corrosion # Tapered, satin finished handle for better two-hand grip and less weight.........

31. Slide out tape measure on hand tool handles

This idea integrates a sliding mechanism in hand tool handles that carry a tape rule inside...The slide out tape rule can be a bit rigid...Now the users can measure a few inches off the handles in there hand tools...A positive feature cause a lot of times we forget the darn tape measure.....Reference data: Product Features:.. # Slide button mechanism in handle for quick, fast adjustment # DIY, industrial or professional use # 15/16" and 1.2" jaw capacity works with various fastener sizes # Debris shield protects slide mechanism, and maintains a smooth operation # Forged steel # Chrome finsh resists rust and corrosion # Polished head # SAE/ Metric size indicator on jaw.......

32. Ratchet socket wrench

This idea designs a ratchet socket wrench exactly like the one described below except is adds a light and magnetized gripper for detailed work....Reference data to pictured item:... The Crescent R2 RapidRench. The Ultimate in Versatility Never before has there been a tool quite as amazing as the R2 RapidRench™ adjustable ratcheting socket wrench. By combining the versatility of a standard adjustable wrench with the speed of a ratcheting socket set, the R2 RapidRench has capabilities beyond anything previously available. Add the power of the famous Crescent® brand, and you have an innovative new product that's a proven consumer favorite.........

33. Hand warmer integrated into hand tool handles

This idea integrates a hand warming system into the handles of hand tools such as screwdrivers, hammers, saws and even tape measures....The product is a zippo lighter with a hand warmer. This is the perfect benchmark...Since most of us work with our tools outside or in rather cold conditions, it would be nice to have a tool that also warms your hands....Some people don't use work gloves with tools....Respectfully.....Reference data on pictured item..:Zippo HAND WARMER - Z20088...Item Description: Enter the charismatic Zippo Z20088 hand-warmer you can take anywhere. This delightful pocket friend can keep your hands warm for up to 24 hours, and its even 10 times hotter than ordinary hand-warmers, thanks the the original premium lighter fluid used. Zippo enthusiasts will be beaming when they see this, because there is such a strong bond between Zippo collectors and the products. This particular set comes complete with 4 oz of lighter fluid, a warmer bag and the usual high-spec presentation box. What are you waiting for – its cold out there!...........

34. Motorized hand truck for contractor and Do it yourselfers (DIY)....

This idea designs an economical hand truck with motorized movements like the existing truck called "ULTRLIFT1500"..My idea benchmarks to this lift truck except reduces the cost for home and small contractor use...I got to see this truck in action and it's amazing....Please see the specifications below as the perfect benchmark....Respectfully.....Reference data:.... reference :... Item # Ultra Lift Model 1500 Powered Hand Truck Customize Your Own We listen to our customers - ULTRA LIFT Powered Hand Trucks and accessories have been developed to meet the unique requirements of the markets we serve. The Patented ULTRA LIFT is fabricated of custom aircraft aluminum extrusions. Machine weight includes strap bar, 4 wheel dolly and spill-proof sealed battery. Minimum over-all machine height is with extension handle retracted for moves in restricted areas. Maximum height is with handle fully extended for maximum leverage. Call for information on additional models. Many specialty application models, custom units, and accessories are available. ALL MODELS ARE SHIPPED COMPLETE WITH: Variable Position Strap Bar w/ Nylon Strap Extension Handle to Adjust Overall Machine Height Swing Out 4 Wheel Dolly Attachment Leverage Bar for Easy Break Back of Loads Safety Over-ride Clutch for Drive System Fully Automatic Anti -Reverse Break Spill Proof, Sealed Battery - 12 Volt, 34 Amp Hour Battery Charger with Pre Wired Charging Plug oad Capacity1500 lbs. Lift Height41" Machine Height62 - 72" Machine Weight120 lbs. List Price$3400.00 Accessories Model SEH - Strong Arm Extension Handle ($300.00) Model LPE7 - Lift Plate Extension ($105.00) Model LPE15 - Lift Plate Extension ($105.00) Model CBA - Concave Barrel Attachment ($95.00) Model VGH - Vacuum Grip Instant Handle ($125.00) Model LPS - Load Positioning System ($350.00) Model LBB - Load Balance Box ($200.00) Model VMA - Vertical Move Assembly ($325.00) Model VMA-S - VMA Adapter Kit for Snack Vendors ($75.00) Model FHA - File Handling Attachment ($225.00) Model HMA - Horizontal Move Assembly ($250.00) Model PLE - Portable Landing Extender ($475.00) Model PLE/LP6WD - Combines PLE w/ Low Profile 6 Wheel Dolly ($825.00) Model CGM - Case Goods Mover ($175.00) Model ESP - Elevator Swivel Plate ($225.00) Model RTA - Rough Terrain Attachment ($250.00) Model THB - Truckbed Hook Box ($95.00) Model SCL - Strap & Channel Loading System ($95.00)............

35. Hammer with rectangle head

This idea ergonomically designs a hammer with a long rectangle head...The geometry side view would look like the attached picture except the razor edge on the axe is replaced with a rectangle hammer head...Now the user has more surface area upon which to hammer items including nails....The golden ration on this design is good.....Reference data..:Gerber back paxe Forged Swedish steel head. Superior edge retention. Polymid Fiberglass handle molded around head, it cannot come off. Ballistic sheath. Length: 8.875".......

36. Vehicle car seat for storing tools

This idea ergonomically designs a toolbox which is designed like a child car seat for vehicles...This toolbox is portable and meant to sit without moving inside the typical vehicle seat...Let's face it not everybody owns a truck with full blown toolbox in the cab....If designed correctly this toolbox can be very handy.....

37. Bow saw with arc blade

This idea designs a bow saw similar to the one described except the blade is in an arc at 180 degrees....I feel the arc would cause a circular cutting motion and in so doing allowing the user to apply more pressure at a given area along that arc.... This saw would also allow you to use the teeth more evenly and at stronger points....Reference:...15" Bow Saw Model No. 836C815T • Cuts quickly through wood and plastic • Sturdy 15" blade • Comfortable PVC-padded handle • Protective sheath for blade......

38. Fold out tools with spring clip

This idea designs a set of hand tools that fold open and are carried on a belt with a spring clip that detaches...For example, a hammer, a crescent wrench, screwdriver, hand saw.... They could all fold out and be easily carried around...Even a tape measure which is ultra thin can be carried this way..... Reference data: Coleman pocket knife....Pocket Knife Model No.

836A410T • Stainless steel blade • Leather case • Spring-loaded clip................

39. Shop scissor with more functions

Please see Coleman camping scissor as perfect benchmark for this idea....My idea ergonomically designs a shop scissor with more functions integrated into its design including a small vise like gripper.....Similar to the functions stated below and pictured...Conventional shop scissors look so plain Jane that it would be nice if this somewhat large tool had more functions... Respectfully....Reference data:...• 12 functions in one tool:.. * scissors * screwdriver * magnet * can opener * wrench * fish scaler * nut cracker * jar wrench * wire stripper * wire cutter * bottle opener * knife............

40. Tape measure with integrated utility knife

This idea takes the Tape measure like the 33-725 25-Foot Fat Max and ergonomically integrates a utility knife right under the reel of the tape...Now the users have a utility knife right at hand once they measure...These two tools are in the same family and there should be an attempt to integrate them....

41. Hand tools that fold up into cubes

This idea ergonomically designs hand tools which fold up into a cubes. With this idea the cubes are stored side by side for an exacting storage methodology...The extend fold out tool follows the form follows function rule except is add material and hinging points to make these tool fold up neatly....

42. Screwdriver with tape rule inside handle

This idea ergonomically designs a tape rule mechanism inside the screwdriver handle...The tape rule end comes out of the bottom side of the handle facing away from the metal tip...Now users have a convenient tape rule integrated in a screwdriver... these tools are in the same family and could be combined this way...

43. Flip down write pad integrated in tape rule wall

This idea ergonomically adds a second wall to the side of the typical tape rule in order for it to flip down at users will. Once flipped down the user has an erasable note pad inside the wall in which to write down data....When user is done they snap this wall up again to lock it...I believe if integrated ergonomically this idea would be an added asset to the tape rules....The flip side to this idea is to have the erasable note pad just plain on the side of the tape rule...

44. Tape rule w/ pencil write & erase on reel

This idea is to offer a tape measure with a reel that can be written upon with pencil, afterwards the user can erase the markings...Let's face it we all love to mark on the tape measures with sharpies...Why not offer an actual reel with such a feature?...

45. Updated concept – Heavy Tools Helping hand

This idea ergonomically designs a tool to help users stabilize loads and as a third hand or helping hand...Users can reset heavy power tools on this tool as they work....Users use this tool as a stabilizer in all kinds of ways as described below...My version would also be telescoping for height variations........Reference data... 3-H 3HAND FastCap 3rd Hand HD The 3rd Hand HD is the help you have been looking for. If you ever wished you had an extra hand... here it is. The 3rd Hand HD and Little Hand HD provide support, brace, or clamp for whenever you might need an extra hand. The 3rd Hand HD is like an extra person. Great for: * Trim and crown molding: use the 3rd Hand HD to secure your trim or crown molding in place while you nail it in place in another location * Glued down flooring apply pressure to the flooring material while it dries * Drywall: hold the drywall in place overhead while you secure it * Load stabilizer: mount the 3rd Hand HD in place horizontally to prevent material from shifting in transport........

46. Nest to hold a few nails & screws on power tools

This idea ergonomically designs a nest on typical power tool where nails and screws can be attached...This nest can be integrated in the tool shell design or as an attachment, whatever is cheaper. Weather by magnets or otherwise the nails and screws would be right there for the user, in small quantities of course but great for the quick jobs most of us to around the house.... The flip side to this idea is of course a wristband worn nest to hold a few nails or screws, wrist fit by velcro..Benchmark of this idea is to attached picture and following description:... ""
PROHOLD FastCap ProHold...Always loosing your screws? With 9 neodymium high performance magnets, and a convenient wrist strap, you will never be without a screw again. Features of the ProHold: # High performance NEODYMIUM magnets hold screws or nails in place # Adjustable wrist band fits around your wrist, but also enables the ProHold to fit other places, such as the end of a drill # Heavy duty Velco® straps are made to last and last # Flexible magnet surface adjust to the shape you need........

47. Bench vise with ratchet tightening system

This idea ergonomically designs a ratchet locking system into a conventional bench vise....Now the user would actually have powerful clamping holding pressure in a bench vise at the jaws ...Rather than a turn by hand handle the users can easily ratchet tighten in there object...This is great for older retired craftsmen who are not as strong as they used to be....Respectfully..... Reference data....."" The quick and safe method to generating powerful clamping pressure. The new lever clamping system with the Synchro-lock mechanism for easy and conformable operation. * You have the choice to either use the lateral ratchet release lever or the push-button release at the end of the handle. Choose the right press, and click! * The improved leverage and ergonomically designed quick release lever reduce expenditure of time and energy to a minimum. * The new ratchet protection shell avoids injuries of finger when clamping overhead. * The fixed and moveable jaws are made from high quality special steel, drop forged, hardened and galvanized. * The ratchet is wear-resistant designed. * The handle is ergonomically designed and plastic coated. * Large tilting pressure plate, is galvanized...

48. Suction cup w/ pole handle for lifting light loads

This idea designs a suction cup with a vertical pole handle. To clarify that's one suction cup with a pole handle...Would look something like a baseball bat except with a suction cup at the bottom...At the lip of a lever the suction cup holds itself to items as conventional cups......Now the users can lift small loads up top 50 pounds ergonomically without hurting there backs bending over...Respectfully.....Reference data of pictured item:..."" HOD-DOUBLE FastCap Handle on Demand Double # Stop straining your back trying to lift heavy stuff... suck it up! With FastCap HOD, (Handle on Demand) double, just fasten the suction pads to any non-porous surface and move it, lift it, place it, transport it... you decide # 200 pound capacity.............

49. Left handed Tape Rules

This idea is to offer left handed tapes on their Tape Rules like the model 33-599 - 25' x 1" MaxSteel® Tape Rule.....Reference * Right hand (RH) tapes begin at point "0", then ascend toward the right hand in measurement... * Left hand (LH) tapes begin at point "0", then ascend toward the left hand in measurement...I note a lot of us lefties like things reversed...

50. Masking tape with inch mark increments on tape

This idea is to offer a roll of masking tape with inch mark increments printed on the tape...This would be useful for all types of work and reducing the need to use tape measures.... For example people needing to mark nail spacing for one center nailing...

51. Extra wide masking tape with fluorescent edges

This idea designs an extra wide masking tape roll used for all types of work like painting, indicating pathway's, and in the dark too...The edges are meant to stand out day and night...I'm thinking this roll can be six to nine inches wide...

52. Portable digitizer for construction & DIY projects

This idea ergonomically designs a hand held digitizer for the construction fields...The user would set a zero point on this device using ground positioning satellite technology (GPS)...Next the user begins to "touch" important points along whatever path of measurement is chosen and needed.....The result would be a digital map on the screen of this device with exact measurements...For example if i wanted to get the geometric measurements of the perimeter of a house...This device can also do roofing geometries...Perhaps some surveying as well.....Respectfully.... Background of a digitizer:..."""" Digitizing or digitization[1] is representing an object, image, document or a signal (usually an analog signal) by a discrete set of its points or samples. The result is called "digital representation" or, more specifically, a "digital image", for the object, and "digital form", for the signal. Analog signals are continuously variable, both in the number of possible values of the signal at a given time, as well as in the number of points in the signal in a given period of time. However, digital signals are discrete in both of those respects, and so a digitization can only ever be an approximation of the signal it represents. A digital signal may be represented by a sequence of integers. Digitization is performed by reading an analog signal A, and, at regular time intervals (sampling frequency), representing the value of A at that point by an integer. Each such reading is called a sample. A series of integers can be transformed back into an analog signal that approximates the original analog signal. Such a transformation is called DA conversion. There are two factors determining how close such an approximation to an analog signal A a digitization D can be, namely the sampling rate and the number of bits used to represent the integers....

53. Conventional portable ladder with elevator lift

This idea designs a ladder like the one pictured except there are no treads instead a lift who's geometry looks like the stand geometry pictured, with the man standing on it...At the flip of a cord switch by the users side the mechanism goes up and down...Two vertical legs on the ladder can have a rack design and of course the pinion gearing on the stand...Now the users have a firm place to stand and exact height adjustments for there work...The key to this product is for it to be lightweight....

54. Forearm forklift for construction trades..

This idea benchmarks to the company referenced below:...This idea is similar except for the construction trades.... Respectfully:..""". Above All Co. The Forearm Forklift moving straps were designed in 1997 by a professional mover who is still very active in the industry. After many years "on the truck" he felt compelled to invent a tool that actually eliminated the risk of floor damage. A claim that only the Forearm Forklift can make since the dolly and hand truck require the rolling of wheels on your sensitive floors. Coincidentally, he also designed them ergonomically to encourage proper lifting techniques and body mechanics. They are actually the first and only OSHA accepted moving tool ever to reduce potential injuries due to heavy and repetitive lifting. * A pair of patented straps that employ leverage * They make anything you carry seem 66% lighter * They encourage proper lifting techniques * They are great for keeping wheeled apparatus' off of your floors which keeps them from getting damaged * Each pack comes with 2 straps which are a complete set * Each pack comes with easy-to-use and illustrated instructions * Each of its 2 straps are 9' 4" long and 3" wide * Rated for carrying up to 600 lb (272 kg) * Each of its 2 straps are adjustable up to 48 inches(122 cm) so they' re great for carrying small, medium, large and extra large pieces * Just cross the straps underneath the item that you' re carrying and they never slip.

55. Stilts with golf type shoe pads & height adjust

This idea benchmarks to the ""Marshalltown 14901 SkyWalker® 2.0 Stilts 24-40" (SW224)....This idea is similar to the walking shoe pads with spikes like golf shoes...With spiked shoes the user has a more firm grip on the ground...This idea also gives the user adjustable height variation.....Now the painter as well as the framers, nailers can grow just that little bit as to not use a ladder......Respectfully....Reference on stilts:.."" The advanced ankle movement allows a more natural walking motion. This allows new users to more easily get adjusted to working in the stilts and experienced stilt users will find these to be easier with which to work. Add the cushioned calf pad and you have one of the most comfortable and least tiring drywall stilts on the market today. * Adjustable ankle spring allows a natural walking motion which lets the user stand up straight making work easier on their back * Three-way setting ratchet system to fit any size leg * Durable quick adjust-and-release straps * Flip-and-switch system allows for quick height adjustments up to 40" * 225 lb. load limit * Rubber sole allows for longer life and better traction * Individual replacement parts are available to extend the overall life of the stilts.

56. Portable air tank for pneumatic nailers

This idea benchmarks to the "Marshalltown E400 The Enforcer™ Portable Texture Sprayer (10400)"...This idea is for a portable air tank to use with pneumatic nailers...I believe it would be nice to offer this option...The second idea on this is to have a smaller system and have it battery powered with the battery being detached from the carry air tank....Respectfully... Reference:.." The 2.5 gallon tank capacity allows you to hold more texturing mud then a standard hoper # With a flow rate of up to 1.75 gpm you get greater control over the amount of mud released # Utilizes one air hose that can be used with any standard air compressor # The Compressor does all the work for the pressurized tank and the hopper gun # The piston seal aids in cleaning the inner tank wall as the mud is used up eliminating unneeded mess when the job is finished # The entire unit weighs

15 pounds empty and 45 pounds full # Uses a maximum of 15 PSI in the tank no matter what the compressor is set at for the hopper gun # The quick-release top cap with pressure release pin easily removes for easy cleaning and filling # You can maximize your efficiency by using the 10410 mud pump when refilling # Has four texture settings including knockdown, popcorn, orange peel and splatter # Prior to each use of the ENFORCER™ check all connections and check for worn parts # Use only official ENFORCER™ replacement parts for maximum product life # Made in the U.S.A......

57. Telescoping support benchmarks to maulstick

A maulstick has a ball-shaped end covered with fabric or skin; it rests on the easel and is used to support the brush hand.... My idea is similar except to use for tools which need pinpoint accuracy and are hand held...In other words this idea is a telescoping rest for all types of tools needing hand precision, such as drills.....

58. Bench vise with tightening swing arm

This idea is to offer a conventional bench vise except it would have a swing arm for tightening the jaws. I believe a conventional arm would add a small touch of ergonomics to the bench vises...I have always hated the little steel bar used on vises currently to tighten the jaws.....

59. Craftsman style Riffled file

This idea designs a round file with the geometrics of the riffled file...I believe there is a place in the market for a curved round file similar to the riffled except larger as used in the woodworking industry.....One blade could be for rough and the other for smooth filing.....Respectfully.....Reference on rifler file:..."Riffler Files are indispensable for cleaning up after carving tools and router project. The selection of angles and curves fit into just about any slot or crevasse you need to clean out. These fine cut riffler files are used for finished smoothing. Profiles include flat, round, curved, square, triangle and other associated shapes. 10 pieces 5-1/2" Long….

60. Flat and round file with flexible movement blade

This idea ergonomically designs a flat and round file as pictured except the blades would be somewhat flexible...I believe the craftsman would appreciate a somewhat flexing blade to use on rolling round material and conditions.....Now more artistic finesse work can be accomplished by hand filing...Exact flex angles to be determined...Respectfully....Reference.."Hand tool made up of a metal blade whose striated surface allows for pieces of wood, metal or plastic to be smoothed, altered or burnished.....

61. Protractor w/movement arm (laser & digital read)

This idea designs a half circle protractor with a movement arm movement from zero to 180 degrees. This swing arm is pinned at the center of the half circle protractor which allows swinging.......Now the user has a protractor with a movement arm and laser attached....Easy reading angle measures with a digital readout integrated into the protractor design...This tool uses batteries also....Respectfully....Reference material:.....77-198 - S2XL FatMax™ Laser Level Square Click to Enlarge Click to Enlarge Features and Benefits * Projects horizontal and vertical laser "chalklines" on surfaces and wall * Ideal for floor and wall tiling, paneling, wall fixtures, wallpapering, stenciling, wainscoting, finish carpentry, outlet switches, lighting fixtures, wall studs, partitions and much more * Non-abrasive vertical mounting - push button application with adhesive putty mounts instrument to surface * Aluminum Precision Machined 90° edge * Includes Target, Adhesive putty (2), Carrying Case, Batteries, Users Manual * 4X brighter beam Line Accuracy - 1/4 at 60 feet Working Range - Up to 100 feet depending on illumination of area Laser Diodes - Two (2) 635nm diodes Power - 3 "AA" batteries (included) Waterproof (IP 55).....

62. Gyroscopes with digital readouts on Fatmax levels

This idea designs a new level with a horizontal, vertical and angular gyroscope...Each gyro to have a digital readout of angle measure...This idea basically replaces your liquid tubes on your levels and replaces them with gyroscopes , each with there respective digital readout...This tool could have an integrated battery system to run electronics.....Now the user plays no guessing games and gets exact angle measurements......

63. Telescoping shanks on screwdrivers

This idea ergonomically adds a telescoping shank to the typical screwdrivers...The reason for this is storage...Now the shank can telescope inside the handle or with a small protrusion from the handle...Users can then store many more object in tool boxes and such, plus for easy toting...We make sure the telescoping mechanism inside handle for this idea has a backlash feature to make sure telescope shank out is secured...

64. Cordless electric table saw with wheels

This idea ergonomically replaces the electrical cord with a detachable rechargeable battery to operate a small portable table saw...This saw will also have a suitcase type handle and wheel for easy portability....The key here is the ergonomically designed detachable battery. Now the user has a heaver cutting power tool plus the portability easily tote..

65. Digital angle readout on hand mitre saw

This idea adds a digital readout for angular position as related to the hand saw...Now the user plays no exact guessing games with small increments by visually seeing the angle position with big digital number readouts located ergonomically on the mitre scale area of the tool pictured.....

66. Torque bar thru crescent wrench handle hole

This idea designs a metal bar which inserts through the hole at the crescent wrench handle....With a bar through the hole the consumer now has both hands to grasp the bar at each side of the wrench handle and create a stronger torque to tighten or loosen a hardware item like a nut.....

67. Motorize worktable for slow movement

This idea ergonomically adds a small motor and gear train in order to add a slow movement when the worktable table is carrying some heavy load...I note it happens to me a lot when I have a heavy load on my worktable and need to move it...I believe by adding a slow moving gear train it will enhance this table........Reference on the Fatmax table..."" FAT-MAX MOBILE PROJECT CENTER * Portable work center, clamp table & hand cart * Electric power: 3 sockets & cable holder * Work surface: 660 lbs. load capacity * Hand cart: carry 220 lbs. * Telescopic cart handles * Versatile clamping w/multiple peg holes * Easily folds flat for storage Mobile Project Center Unit: 1 Each Manufacturer's #: 93-292 Ship Weight: 44.00 lbs Size: 22-3/4" Height Stock Status: In Stock Stock detail Suggested Retail: $ 149.99 Our Price: $ 116.09....

68. Sugar removing filter for sodas into cups and glasses

This idea designs a filter which is hand fitted over any standard drinking cup or glass...This filter is used when the user wants to cut down on there sugar intake...The user pours there juice or drink into there given cup or glass and this filter traps as much sugar as possible.....I would love this idea of having my tasty fresh orange juice yet cutting down on the negative sugar side effects in the body...I note athletes would love this product.....

69. Slow-down shot mechanism on pneumatic nailers

This idea designs a slow down mechanism inside your pneumatic nailing gun....The user can option to slow down the nail shot speed to prevent shooting themselves by accident.....I believe the consumer has a right to flip a switch on your pneumatic guns to slow down the shot speed.....I believe safety comes first, side note my dad worked at the Naval Shipyards in Long Beach California for 30 years with no major accidents, his number one rule was safety not speed...A few seconds extra per shots, who cares....I note a great safety feature here....The physics behind a slow motion shot mechanism is possible with Douglas fir type woods used in construction...Now the user has time enough to pull a body part away from the proximity of a nailing hardware point.....

70. Handheld microwave drying tool for lumber sites

This idea ergonomically designs a hand held electric powered microwave drying tool to use when wood construction jobs have been rained on....This tool uses the correct microwave bandwidth to ensure it's microwaves penetrate solid wood items such as studs and beams...I believe this tool would save contractor time in quickly drying off job wood building structures....Moisture content in lumber over 19 percent is not allowed in wood framed buildings...Normally contractors wait a few days before returning to job sites after rain...Why not offer a tool which is hand held and is a sweeper in design which they pass along there wood structure in order to dry off the lumber quickly?....Respectfully.... This idea benchmarks to my previous idea number 1706 as referenced here:...."" Date 23-Aug-2008 ID 1706 Subject Microwave oven to dry lumber (studs,rafters etc) Description This idea ergonomically designs a microwave oven which can hold wood pieces such as studs, rafter, floor joists etc...This oven has the same concept as your home microwave but the geometry is different in order to fit lumber sizes...The exact temperatures to be determined...Now the contractor has a tool to use in order to dry and cure wood quickly instead of air dry...If this idea takes off it can be used at the lumber clearing houses to also dry amounts of lumber quickly....This tool can be sold in different sizes to the size of the company it serves...

71. Tools w/ hollow handles for loctitie & oil storage

This idea ergonomically designs small storage tanks inside certain tools used for tightening or loosening hardware items such as bolts and screws...As example, the attached pliers could have hollow handles. One handle to store loctite , the other handle to store lubricating oil....A small mechanism can act as the open and close tip on these handle storage tanks....Now the user has there needed tool plus the liquid necessary for that job.....

72. Cordless electric router w/advancing mechanism

This idea ergonomically designs a cordless electric router with an advancing mechanism...I believe a cordless electric router would fit perfectly for today's on the go craftsman...The advancing mechanism on this router benchmarks to vacuum cleaner advancing mechanisms...I believe an 18 volt rechargeable battery is the correct size for this tool...This tool to also have a wall recharging mount with the option of recharging from vehicle cigarette lighter.....Now we have a portable use anywhere router...Why not compete with Craftsman tools?....

73. Chalk line tape measure tool

This idea ergonomically designs a chalk line tool as pictured except the string is thicker and has incremental measurements like a tape measure...There is no chalk on this tool it is meant strictly for measuring contours and straight lines as a chalk line would be used.......Reference material on conventional chalk tool:...""""chalk line.. Instrument consisting of a cord that rewinds into a case filled with chalk powder; it is used for marking straight lines. case click to hear Metal body housing the chalk powder and the line. hook click to hear Curved metal end of the cord that can be attached to an object; it also makes unwinding easier. line click to hear Chalk-covered cord that is pulled between two points to mark a straight line. crank handle click to hear Device for rewinding the line into the case containing the chalk powder.....

74. Portable machine that makes washer geometries

This idea ergonomically designs a small possibly hand held sort of press brake punch for making different geometries of washers....This new machine gives the users the ability to produce small quantities of washers for specific uses...I note an overall savings here if this machine could be made cheap enough....Please refer to the picture as example of four differ-ent washers this new machine could produce from a a single "generic washer" ...These generic washers are used as the base upon which to contour whatever washer geometry is needed... The user would place this generic washer inside this machine then pick and choose geometry of washer to produce as many as desired...Respectfully...Reference material:..Ringlike parts placed between a nut or a bolt and a part to be tightened; they distribute the stress.....

75. Automatic square or round file

This idea designs what looks like a conventional file with a handle...The difference in this new design is that at a press of a button on the handle the square metal file area can bend up-wards in a circle fashion to create a round file..A small mecha-nism inside the handle and bendable metal creates this space age tool..Now the users have the best of the typical square and round file in one tool.......Definition of the conventional file ref-erenced above:....""..Hand tool made up of a metal blade whose striated surface allows for pieces of wood, metal or plastic to be smoothed, altered or burnished""".

76. Flashlight with optical comparator lens

This idea ergonomically designs a special flashlight with an optical comparator lens which the user can light upon a typical wall then go measure some item against the wall....This flashlight has a laser switch and readout which tells the user how far from the wall to be in order to get the correct scale measurements...

Now users have a sort of tape measure comparator with added angles to measure 360 degrees...This new tool flashes a circles with degree measures as well as incremental measure marks in whatever scale is determined....Respectfully....Reference material on Optical comparators....The old adage "seeing is believing" is appropriate when referring to optical comparators. Because these measurement tools display a magnified image of a part, a tremendous amount of information about that part can be gathered in a short time simply by looking at the image.... Optical comparators, for those unfamiliar with them, are inspection machines that project magnified images of parts onto a glass screen using illumination sources, lenses and mirrors for the primary purpose of making 2-D measurements. ... Dating back as far as calipers and micrometers, optical comparators have been used for more than 50 years and remain a versatile and cost-effective technology for monitoring the processes and quality of a broad range of manufactured parts. Originating from static overhead projectors that displayed magnified images of screw threads onto a wall for manual measurement, optical comparators have evolved into full-featured machines that use modern mechanical, electrical and optical technology to minimize inspection time and maximize cost savings....Comparator advantages.... Optical comparators can provide more information than just simple dimensions. Length and width measurements of the part shown above, for example, can be quickly obtained from two separate measurements by using a micrometer. These superficial measurements, however, might not reveal burrs, scratches, indentations or undesirable chamfers. Such imperfections are best detected on a comparator. In addition, a comparator's screen can be simultaneously viewed by more than one person and provide a medium for discussion, just as a white board might facilitate a conference......

77. Wall mount organizer and battery charger

This idea ergonomically designs a wall mount organizer and battery charger to all future battery powered tools which will offer to the consumer....I note this might not sound like a big deal but it is....This idea benchmarks to the "WORKSGT 2 in 1 Trimmer/

edger" They have designed a special wall mount and battery charger...""" Question: My battery will not stay charged very long, why not? Answer: Each battery pack comes UNCHARGED so you must charge the battery before use. Your battery will not reach its full capacity unitl you have charged and discharged the battery pack several times. FOR OPTIMAL PERFORMANCE OF YOUR BATTERY, AS YOU USE YOUR TRIMMER FOR THE FIRST 3 TO 4 TIMES, IT IS RECOMMENDED TO FULLY DISCHARGE AND FULLY RECHARGE THE BATTERY PACK. This will ensure that you get maximum runtime from your battery pack. If you require longer working time, you can order additional battery packs.

78. Tools with integral ROM chip voice instruction

This idea ergonomically adds miniature digital ROM chips integrated into tools so the typical user can listen to instructions on use and safety...The voice instructions are amplified thru small speakers onboard the given tool...This idea benchmarks to musical greetings cards in which the user opens the card and some music is played...I believe the users would appreciate onboard instruction on use and safety...The flip side to this idea is in fact the same idea as stated above except in an open card type design as the greeting cards on the market today....Example. The user would open there tool packaging and listen to the tools instructions when they open the tool use card...

79. Suction cups on tools

This idea ergonomically adds integrated suction cups into hands and other ergonomic places on hand and power tools... The purpose is to be able to secure tools to flat surfaces to prevent them from vibrating around and falling....

80. Ergonomic keypad speed governor on vehicles

This idea simply adds an on board touch button screen or keypad in which the consumer can quickly type in there speed limit maximum and let the vehicle cruise at that speed....The purpose is to save on gas and money....I note a lot of gasoline is saved when we drive 55 miles per hours over say 70 miles per hour...Why not offer an add on product which the consumer can have installed at the vehicle dealerships?....Current cruise controls are rather hard to figure out and not functional for quick adjustments...For example, I am on a road and want the car to cruise at a max of 35 miles per hour...next block 45 miles per hour, then I enter the highway freeway and want to cruise at 55 miles per hour...I would love an ergonomic tool which let's me quickly and easily be adjusting my cruise maximum speed.....I hope my explanation makes sense....

81. Rasp with hacksaw body

A rasp is a hand tool made up of a metal blade whose tooth covered surface can quickly rough out wood, metal and plastics....This idea ergonomically adds a hacksaw type body over the rasp handle to end for better handing and improved power control when rasping.....The hacksaw body could be disconnected to leave the rasp only as pictured ....

82. Lightweight square pipe clamp w/rack & pinion

This idea designs a pipe clamp like the one pictured except with a lightweight metal square tube with rack and pinion movement....I believe the conventional pipe clamp is to heavy with insecure movement...With the new design we have lightweight feature plus a pinion movement on the jaw and tail stop... The square tube would have the actual rack upon which the pinions move...There would be back lasj mechanisms on both upper and bottom clamps...

83. Electricians hammer w/pull out fish wire in handle

This idea adds a fish wire reel inside the electricians hammer which comes out the bottom of the hammer handle...Now the hammer itself is used as a handle for the fish wire work...The fish wire reel winds in and out with a pull out turn knob located along the hammer handle...

84. Wood chisel with opposite side metal blade

This idea ergonomically adds a second chisel blade on the opposite side of the handle.....The handle i made longer which reduces the length of the actual metal blade...The theory is that nobody really sharpens chisels all the way up to the handle and in so we save precious steel by making the metal part shorter.... Now the user has to chisels in one plus its a cost savings....

85. Hacksaw bridge saw integrated on handsaw

This idea ergonomically adds a hacksaw to the typical hand-saw....The handle on this new handsaw would now become reversible...The hacksaw blade would face in the opposite direction as the handsaw teeth...In attached picture the "back" part of the blade would have a bridge over it from "toe" to "heel" to help support the proposed hacksaw mechanism...This would all be very ergonomic....Now the consumer has a 2 in 1 tool for cutting...These saws are in the same family, why not fuse them together....

86. Cordless tools with dual powering ability

This idea simply gives the typical cordless power tool two power sources, battery or conventional power cord...The power cord is attachable via a plug in port on the cordless tool....Now we have two methods of powering our cordless tools...Sometimes it's "either or" condition, either the battery has gone dead or there is no place to plug in the power tool...Let's offer both solutions to the consumers...Side note the power cord attaches and detaches on these tools...

87. Hand held pneumatic tool for demolition work

This idea ergonomically designs a small hand held tool which acts like a super rip bar because this one is pneumatically powered with your compressor...This idea benchmarks to "Holmatro Incorporated 4150 Combi-Tool"..I believe the consumer would appreciate a little bit more help with a tool one notch above the 55-099 - FatMax® Xtreme™ FuBar™ Utility Bar - 18" as example.....

These are the functions of the Holmatro tool which could be incorporated into this new pneuamtic tool.....Respectfully...."":• Multi-functional - cutting, spreading, squeezing and pulling with one tool. • Unique new blade design optimal cutting results, perfect spreading grip. • Deadman's handle. • UL Listed. • CORE Technology™. • Lighted Handles. • i-Bolt technology....

88. Smaller version of "Brace" tool for screw driving

This idea is to ergonomically design a screwdriver which can be turned like the Brace driller pictured in my attachment ,except a smaller version which is ergonomic, maybe 50 percent the size of normal Brace tool......I believe not everybody owns a cordless screwdriver...I believe not everybody wants a cordless screwdriver with battery hassle....I believe the consumer would appreciate an actual screwdriver which can securely tighten down a screw using hand action only...Various screwdriver tips can be stored ergonomically on this new tool....The handle and front knob on this tool to be rich cherry wood...The mechanism refined for movement....

89. Reintroduce Spiral Screwdriver with refinements

This idea is to reintroduce the spiral screwdriver as my attached picture shows except with refinements...The various screwdriver tips could be stored inside an ergonomic handle. The chuck and ratchet to be improved in movement and locking feature....This type of screwdriver has a retro look and feel to it. I used this type of screwdriver before and feel it could make a comeback with refinements...I believe the handle could also be a rich cherry wood larger diameter for better feel....

90. Safety hose which inflates to help fall victims

This idea ergonomically designs a safety tool which looks like a garden hose except this hose expands when is inflated by air pressure and expands in a curled up fashion along a given path.....The user would lay out this hose around the perimeter upon which a human will be working from some altitude...If that human was to somehow fall down by accident then this inflated hose would somewhat help that fall and at least help to prevent a death....I notice all construction sites use high scaffolding...The planking used on scaffolding is very loose since it is not nailed down...Why not give the contractor a safety tool which helps make the worker feel safer?.....This product could be sold like a garden hose with varying lengths being offered.....

91. Handsaw with measure twice cut once laser pointer

This idea ergonomically adds a miniature laser pointer on the top portion of the "toe" part of the hand saw which is pictured in my attachment....I believe if the user had a visual laser pointer as they are sawing they would definitely cut in a straight line plus help them not over cut....I believe this laser pointer would also be a heads up sort of tool in which it would indeed help the user measure twice before cutting something expensive....

92. The flexible handle crescent wrench

This idea ergonomically designs a typical crescent wrench except the handle is somewhat flexible...Now the users can get into tight places and actually do some bending laterally....This new handle can get into tight places...Since a handle is still a handle and acts as a torque then that item can still be tightened or loosened.......Exact material to use for flexible handle to be determined...The crescent mechanism can remain as is in hard steel....

93. Body harness to rest heavy tools

This idea ergonomically designs a comfortable body harness to help carry heavy work tools such as the pneumatic nailing guns...This harness would have pockets all around the harness to temporarily rest the heavy tool as the user works...When kneeling this harness also has a pocket around the lower leg area so the user can rest there tool.....I'm still noticing the guys who work with heavy tools get exhausted after a days work. Let's help..

94. Tool belt for hand saws

This idea simply design a special tool belt to carry hand saws...This belt has long sheaths clipped onto it which protect the hand saws from damage as they are worn...The hacksaw, handsaw, compass saw and coping saw could be the ones worn on this belt.....Now when toting hand saws around they can be protected and easily retrieved.....

95. Magnetic hook on tape measure

This idea simply adds a magnetic hook on the typical tape measure...I note a lot of tapes are used in measuring metals as well...Why not magnetize the hook for better holding power....

96. Glow in the dark paint sheen on hammer heads

This idea ergonomically adds a glow in the dark paint sheen which does not distract from seeing the metal hammer head... This sheen has to be light yet visible to people nearby...The reason for the this safety paint on the hammer head is of course to mark the arc of travel as the user is hammering away...I believe this would be great when the sun is going down and when large groups of people are around...Imagine the claw hammer pictured (attached)...That one could see the metal yet when the hammer is swinging it could sort of glow to indicate arc of travel.....

97. Safety glasses with second earpiece for ear plugs

This idea ergonomically adds a second ear piece set on the typical safety glasses which secures the glasses to the human face....These new ear pieces are flexible and have holders for the disposable earplugs....Now the user puts on there safety glasses and simply adjusts the second earpieces to the left and right ears ....When finished the user throws away the ear plugs. next time they re fit new ear plugs to the holders on the glasses. I believe these two safety products go hand in hand and so why not combine them...I am always losing my earplugs...Let's fuse them together,,,I hope this makes sense otherwise I can make a sketch...

98. Dual adjusting wood planes for dual line planning

This idea ergonomically designs a set of adjustable hand planes which can be adjusted via a ruler that attaches both planes...Now the user has a way of planning two surfaces at certain distances while keeping a straight line and not interfering in the middle path...Please see attached picture as perfect reference...Respectfully....Reference date to pictured item:..."""If you're trying to look nice but without running to the barber all of the time you might pick up one of these razors. You just adjust it to the perfect width of side burns and you get straight lines. That is unless you happen to jump or something and end up with a small squiggle in the wrong direction. However, as long as you take your time and don't rush through the job you should be fine. I suppose it also could be used in other areas of your beard if you're the type that likes to get creative though. Sadly, this is still just a design but as much as the razor companies are always looking for an edge over the competition I don't see it staying that way. They are always adding more blades and claiming to be far superior of anyone else. The design was created by James McAdam who is from London. Which also could mean that those in that part of the world would get it first and those of us in the US would have to wait a little longer.

99. Ecology friendly plastics for tool components

This idea is to use OX-Biodegradable plastics in tool components across the board in order to advance the Green Earth movement...I have no affiliation with this company but recommend them for my idea....Please see reference data below and there website..."http://biogreenproducts.biz/whyoxo.html""". ."""""Oxo-biodegradable plastics to the rescue"""".....""""Since young I was always taught to reduce the use of plastics and Styrofoam containers/cups as those are non-biodegradable, not to mention that turtles often mistake them for jellyfish only to be conned into swallowing plastic bags, leaving them to die shortly thereafter. Biodegradable plastics are the way to go it seems - but that doesn't mean we shouldn't continue living out other green lifestyles. "" Thanks to an additive that slowly breaks down the disposable plastic items that we use, plastic can be biodegraded and become part of the soil that nourishes the plants that keep our planet's atmosphere in balance. The addition of this additive turns ordinary plastic into oxy-biodegradable plastic. BioGreen Products sells disposable items made of oxy-biodegradable plastic. Hopefully oxy-biodegradable plastics will become the norm in the near future for the sake of our planet and our children's future.

100. Quick disconnect power cord for power tools

This idea designs a toggle type connector that can be quickly connected or disconnected from the power tools...With this idea the user does not have to drag the power cord along with the tool...Instead they carry the power cord at whatever length and quickly connect to the tool via a new electrical port....This idea benchmarks to the connecting feature of the ""Cat 6 Shielded Patch Cables, Network Cables:(Connectors: RJ45 8 Conductor Male/Male)"""....This idea is for handiness and longer wear on power cords and tools...

101. Use Bamboo wood for hammer handle & other

This idea ergonomically designs bamboo handles for hammer handles...In my in depth study I find bamboo to be very lightweight and very strong when treated. It is used extensively in Asia for all sorts of building materials....Why not try a hammer handle?...The user would feel less fatigue. The hammer head would be conventional....Bamboo can even be grown square for other tool handle applications...Respectfully....Reference data about bamboo...""When treated, bamboo forms a very hard wood which is both lightweight and exceptionally durable. In tropical climates it is used in elements of house construction, construction scaffolding, as a substitute for steel reinforcing rods in concrete construction, ... Modern companies are also attempting to popularize bamboo flooring made of bamboo pieces steamed, flattened, glued together, finished, and cut. However, bamboo wood is easily infested by wood-boring insects unless treated with wood preservatives or kept very dry (see picture)."""

Bamboo is the fastest growing woody plant in the world. Their accelerated growth rate (up to 3-4 feet/day (1.5-2.0 inches/hr)) is due to a unique rhizome system and is dependent on local soil and climate conditions.""" Besides its use as a construction material, it is also used for fence making, bridges, toilets, walking sticks, canoes, tableware, decorative artwork carving, furniture, chopsticks, food steamers, toys, bicycles, hats, and martial arts weaponry, including fire arrows, flame throwers and rockets. Also, abaci and various musical instruments"""....Bamboo canes are normally round in cross-section, but square canes can be produced by forcing the young culms to grow through a tube of square cross-section slightly smaller than the culm's natural diameter, thereby constricting the growth to the shape of the tube. Every few days the tube is removed and replaced higher up the fast-growing culm.....

102. Hammer with integrated metal detector

This idea adds a small metal detector to the hammer handle....This metal detector can be integrated or detachable so it does not get damaged when hammering....I believe these two tools go hand in hand and so why not integrate them?...

103. Hot & Sharp tool Mini-Mitts

This idea ergonomically designs small silicone grippers for grapping sharp or hot tools...This idea benchmarks to the store "Crate and Barral" who offer something similar to my idea except for the kitchen...Please see attached picture as reference and there description below........Reference data... Silicone Mini Oven Mitts $9.95 each Crate and Barrel Exclusive Colorful silicone mitts uses gripping "teeth" to grab piping-hot baking dishes and pans....

104. Benchmark pneumatic nailing gun to paint ball guns

This idea challenges a design team to handle the typical paint gun system as a benchmark for a new pneumatic nailing gun...All around the handle and feel and weight and cost of a typical paint ball gun is one hundred percent better than all the conventional pneumatic nailers on the market....This idea is in fact to offer a new nailing gun designed like a paint ball gun system...Please see attached picture as reference in case your not familiar with a paint ball gun....Various manufacturers offer paint ball gun systems including TIPPMAN....Every time I handle a paint ball gun system I can't stop thinking how great it would be to have a pneumatic nailer in this fashion...

105. System of finding your car in a huge parking lot

This idea ergonomically designs a computer system with cameras that are able to read and store each license plate number that enters its parking lot or structure....The license plates are stored in a data base...At user friendly terminals across the given parking lot the users simply type in there license plate and the screen will show them via user friendly pictorials where the heck there car is....My car loses me all the time in parking lots..

106. Tape measure w/integrated range finder

This idea is a typical heavy duty tape measure with the added feature of a laser distance finder....For example the fatmax tape measure with a top portion dedicated to the technology of the FatMax Tru Laser Measurer 100....Now the user has the best of both worlds,,,they measure short distances with there tape measure reel yet are able to turn on the laser distance finder also....

107. Safety belt with gyroscope and air bag deployment

This idea ergonomically designs a safety belt worn around the waist....This belt has air bags tucked around its perimeter....This belt has an integrated gyroscope for vertical position....When turned on this gyroscope will determine if the wearer of this belt is in a fall position.. For example if the person is over a 45 degree point from a vertical standpoint the air bags on the belts perimeter will deploy and so breaking the fall of the user....The uses of this belt are many, from construction to older people....This belt is a bit over sized from a normal belt and so it's meant to be worn over the waist but ergonomic and fashionable...We could see people in the malls wearing these.....

108. Magnetic lumber connectors for construction

This idea ergonomically designs a series of metal connectors like the Simpson brand metal connectors...This idea makes the connectors magnetic in order for them to stick together and not needing any nails or screws...As example when framing a typical home wall why not try to use positive and negative magnetic plates from the stud to the sill and top plates?.....If the magnets could be made cheap then we might have a new system of framing up a building....

109. Hand saw with triangle blade

This idea ergonomically designs a hand saw with a triangle shaped blade with the normal teethed blade on the triangle tip side....the flat side of the triangle has a file for smoothing any splinters from the sawing side....the triangle is not that step but ergonomic....the handle is reversible up and down to use for the file side or the regular saw side....now the users have finally a multi function saw....

110. Hammer handle attachment point at bottom of handle

This idea designs an ergonomic attachment point for all sorts of screwdriver tips...This attachment point is located at the bottom end of the hammer handle....The hammer head would then be used by the human hand to twist and turn....Now the user has a regular hammer plus a tool used to screw down all sorts of items with detachable bits....

111. Two piece hammer handle detaches to reveal saw

This idea ergonomically designs a hammer handle made of two pieces with the seam being along the entire length of the handle. With the human hand holding this handle the bottom part of the handle detaches to reveal a light duty saw attached to the upper handle...Now the user uses the hammer head as the hand grip to perform sawing jobs in light duty....When finished sawing they attach the bottom handle and continue with hammering...A functional tool some folks might like.....

112. Depth probe ruler integrated in hammer handle

This idea ergonomically designs a depth type ruler probe inside the hammer handle...This ruler probe works like that of a vernier caliper as explained below in reference...With the turn of the human thumb on a roller wheel an the hammer handle, the depth probe ruler comes out for exacting measures...This ruler is rigid like that of the vernier caliper...The tolerances to house this ruler inside the hammer are tight for excellent durability....Now the user has a hammer which can also measures depths like that of a vernier caliper...Length of ruler to be designed ergonomically to given hammer handle length....Respectfully...Reference data ""... Parts of a vernier caliper:....... 1. Outside jaws: used to measure external lengths... 2. Inside jaws: used to measure internal lengths... 3. Depth probe: used to measure depths.... 4. Main scale (cm).... 5. Main scale (inch).... 6. Vernier (cm).... 7. Vernier (inch).... 8. Retainer: used to block movable part to allow the easy transferring a measurement .....

113. Hammer with incremental inch markings on handle

This idea ergonomically adds tape measure type markings along the length of the hammer handles (the ones with straight handles)....Now the users have a reference when they are nailing at on center increments and forgot there tape measures... The flip side is to mold into the radius handles a straight edge and put these inch markings....I believe the tap measure and hammer are in the same family group and so why not fuse them together.....

114. Hydraulic shock air chamber for faucets & valves

This idea ergonomically designs an add on air chamber made of a springy copper coil which absorbs shocks when a faucet or valve is closed quickly...This product to be installed without cutting into the typical wall....This product looks like wound up coils atop the faucet or valve.....

115. Hammer with integrated tape measure

This idea ergonomically adds a tape measure which extends out of the handle bottom...The hammer is used as an anchor and lets the user easily pull it out from a fixed location which is the weight of the hammer (if needed of course)....These two tools are in the same circle so why not combine them?....It's totally out of the way of the human hand too.....

116. Hammer with integrated level on handle

This idea ergonomically adds a level to the high end hammer handle, like the ones with the anti vibe handle...I believe these two tools go hand in hand....Why not integrate them?...The key is ergonomics, it would not interfere with hammering nor the human hand.....

117. Hammer with integrated stud sensor

This idea adds a stud sensor to the handle....This stud sensor can be integrated or detachable so it does not get damaged when hammering....I believe these two tools go hand in hand and so why not integrate them?...

118. Accordion power cord for power tools

This idea designs a power cord which is designed like a flexible accordion....Each link in this accordion electrical cord flexes yet remains tight and fixed offering no crazy movements.. Now the user has more control over the crazy power cords used conventionally...If they can use it with the USB why not give it a try with power tools. Now our power cords are tight.....Respectfully... Reference data on pictured item....USB drives are pretty ubiquitous these days and you can get them in some pretty big capacities. In fact, they're so cheap that you can get a gigabyte of storage for $15, and two for as low as $25. Once you get past the 4 GB mark though, prices sky rocket and the selection becomes pathetic. To solve this, a pretty simple and cool solution was concocted: have thumb drives connect together, forming little snakes of thumb drives…..

119. Pest and danger control with new flashlight

This ideas benchmarks to the product described …..Besides acting like a normal flashlight it has sounds which are beyond the human noise spectrum yet affects the animal kingdom in a negative way causing them to run ….Another facet to this new flashlight is an actual repellent it can shoot against human and animals when the danger is closer.....Respectfully....Reference data:..."""It sure takes a whole lot of guts to be part of the riot police as you have to keep the mob at bay without causing serious injury in the process. Thankfully, there have been inventions that make this job easier, and this flashlight is the latest in a long line. Capable of inducing nausea at whoever it flashes at, this works great when you want to disable a target from a safe distance. It utilizes a range finder to measure the distance to the target's eyes, adjusting the energy level of the light to a level that won't cause permanent damage followed by shooting pulses of light rapidly from an array of ultrabright LEDs to incapacitate a person. Sounds a whole lot less painful that a water cannon…..

120. The Smart Projector lays out floor plans on jobsite

This idea is to design a projector in which the basic floor plan of the structure to be built shows up on the actual terrain it will be built…No more prehistoric chalk line measurements to mark off the footprint of the building….This projector is set up in one location of the property and is now ready to display the floor plan….The users would then simply mark off the lines projected onto the property…..This would reduce the time spent by the survey team…Hope this makes sense otherwise I can make a sketch…The projection back to earth method could be from a small flying drone plane from above working in conjunction with this new projector….

121. Affordable home and worksite chemical detector

This idea designs a machine for home and construction site that is able to detect harmful chemicals in the air around the given location...The key would be to make it economical.....Please read my reference data for benchmark...Respectfully...."Hapsite Viper Chemical Identification System"...."Chemical detection ability is extremely important t our troops, and this HAPSITE Viper Chemical Identification System uses infrared technology to improve the identification of toxic substances and chemical warfare agents (CWA's) in the environment in a matter of a few minutes. The Hapsite System that can be used inside a vehicle as well as in open space, is one thousand times more sensitive than NATO requirements for such a system, affording greater safety to troops and civilians in the area. Inficon won an R&D environmental award for the system, just one of several detection devices made by the company.....

122. A positive synergy statement on "GREEN" tool tags

This idea is to attach a positive statement on the Green tool tags of the future GREEN tools....This statement would read (in nice big letters)..."""DOESN'T THIS TOOL MAKE YOU FEEL YOU ARE HELPING TO CONSERVE THIS PLANETS VITAL RESOURCES?... IT DOES US...YOUR A GREAT CRAFTSMAN".....This idea benchmarks to the famous Wrigley's chewing EXTRA bubble gum products...On the inside flap of these little boxes of bubble gum it says:...."""Doesn't bubble gum remind you of your childhood?. It reminds us of your childhood. You were a cute kid".....This makes me smile every time I open the darn box....

123. CO2 Collection filters for household products

This idea is to help fight global warming. A GREEN idea...My idea is to design all sorts of CO_2 gas collection filters which later can be frozen for future disposal into the floor of the deep blue sea...CO_2 gas emissions by humans amount to 30 billion metric tons per year...My idea begins the process of a corporation helping to fight global warming with an actual product like these filters....The filters collect the CO_2 for one year and are disposed of in a green way.....Reference data:...."The average American generates about 7.8 tons of carbon dioxide per year powering their homes and driving their cars. That is about the weight of two full-grown elephants and works out to about 47 pounds of CO_2 a day for every man, woman and child in the country. How do we do it ... and what is a pound of CO_2 anyway?.....It helps to think of that CO_2 as being trapped in a balloon. A one-pound CO_2 balloon would be about 2.5 feet wide. Forty-seven such balloons would fill up your living room—every day. .... examples:..." Dryers typically produce up to 1,446 lbs of CO_2 annually, far more than almost all other household appliances. To find out how much CO_2 your dryer is responsible for emitting, use our calculator...." Refrigerators are the most power-hungry appliance in the home and can account for over 2,800 pounds of CO_2 released annually. Check out how much your refrigerator is emitting and how you can trim off some of those pound....."""" Showers account for approximately 30 percent of household hot water use and can produce over 1,000 pounds of CO_2 per year. To find out how much more CO_2 your shower is responsible for emitting, use our calculator............

124. The flashlight flare

This idea designs a flashlight with the opposite end being a safety flare that flashes forever...Now the user does not need to buy actual hazardous safety flares because they have it all with this product...The rim of the actual flashlight side has a magnet for easy attachment to car in emergencies....Please read below for benchmarking...Reference data:..."" The flameless flare is to the traditional flare what sliced bread is to unsliced bread. It's a LOT better...especially when making a sandwich. Seriously, the flameless flare is an incredibly useful roadside emergency product. The LED bulb will last forever and the magnetic base allows you to stick it to any surface of your car. For people who like: safety, car stuff Features & specs: * Visible from 1,000 feet * Magnetic base adheres to trunk, hood, or door * Doubles as a map light * LED bulb never dies..........

125. Help U.S.military with robotic helmet

This idea is to help our military in foreign countries deal with enemy snipers....I'm not sure if your aware but a sniper can pinpoint a huge bullet on your face exactly where he wants it.... Our soldiers in Iraq have been huge victims of these snipers... In one video i watched an enemy sniper in Iraq took out 25 of our soldiers with exacting shots to there heads..it was not pretty..,..My idea designs a new combat robotic helmet...Laser ultrasonic sensors ergonomically attached to this helmet exteriors detect if any object is moving in its direction at a fast speed.....Ultrasonic sensors use light which faster than that of a fired bullet...When the sensors detect an object approaching like a bullet or shrapnel, the helmets robotics inside jolt the helmet in the opposite direction of the approaching object so the shot will miss the target which is the soldiers head...The jolt itself can be a mechanism ergonomically designed like a strap from helmet to combat uniform, to jolt in different directions...The user which is the soldier is forced to move there head via the robotics inside... This helmet looks like the normal one but adds a lot of flashlight heads all around the exterior, these shine lasers in all directions as the user moves there head..The logistics of this idea is not far fetched , in fact i got the idea from tinkering with the lego mindstorm nxt robotics sets....Ultrasonic sensors measure distance, detects motion, detect object proximity...If any of the minds can add to this idea then i believe it's worth passing along to our Pentagon as a life saving idea for our troops.....

126. GREEN electrical outlet automatically turns off

This idea ergonomically designs an electrical cord outlet which automatically turns of items after a certain amount of time to save power...A green idea....This idea benchmarks to the item pictured and described below except for industrial.....This is a "green" gadget; a multi-way adaptor that acts as a surge protector, but at the same time automatically switches off devices that have been left in standby mode for a pre-determined time.... Most people think that when they put an electrical appliance into standby mode, the only energy used is to light up the red

LED. Searching around the net for material for this article, I was shocked by some of the reported statistics on the energy wasted by devices left in this mode......The first search result I found on a search for "energy consumption gadgets standby" took me to the BBC News website, which had an article on energy wasted in the UK by devices left on standby. A UK survey by the Energy Saving Trust found that the average household had 12 gadgets on standby at any time wasting more than £740 million of electricity each year. It's not just the money that we're wasting; we're also contributing to global warming by needlessly releasing CO_2 into the atmosphere.....Specifically, the estimated CO_2 emissions from gadgets left on standby were:.... • Stereos - 1,600,000 tonnes • Videos - 960,000 tonnes • TVs - 480,000 tonnes • Consoles - 390,000 tonnes • DVD players - 100,000 tonnes • Set-top boxes - 60,000 tonnes ....making a total of 3.6 million tonnes of CO_2 against an estimated total of 150 million tonnes of emissions (carbon equivalents) each year in the UK.......Reliable data for energy leakage in the US were more difficult to come by, but a 1998 study commissioned by the US Department of Energy estimated that 98 million households wasted some 7.7 TWh of electrical energy. This is equivalent to 50 million (metric) tonnes of CO_2, based on my VERY rough calculations from the raw numbers......In any case, wherever in the world you are, you can save a little money and maybe reduce our impact on the environment by unplugging all those gadgets before you go to bed. PCs and monitors are the worst offenders.....You can also buy one of these environmentally-friendly multi-way sockets for $32.........

127. Flashlight with color changing beam

This idea designs a flashlight with various L.E.D.'s inside that can change the color of the beam...The user chooses the color from a special button...Colors add different meanings to flashlight just as street lights do....Respectfully....Reference data:..LED color changing light bulb fits into a standard light bulb socket. Inside the light are four primary colored LED lights which can be combined to make any one of 16 different colors, all of which you can control via a remote control. The light uses only 3 watts of electricity and can do such fancy tricks as flash, strobe, fade, smooth and dimming. That's pretty neat that you can add a dimming bulb without having to install a dimmer switch in your wall because if you're anything like me- doing handy work involving electricity is not a good idea. Costs around $21...

128. Hard dots on finger tips of work gloves

This idea adds small hard round dots to the finger tips of your work gloves...The reason for these dots is so the user can operate small items like cell phones without having to take the gloves off.....

129. The team toolbox

This idea designs an ergonomic team toolbox that stacks... Each person can connect or disconnect there toolbox in this design....Great for space saving needed in shops and industrial work areas...For families too, in there home workshop garages... Nobody produces this idea...Let's try it...Respectfully... Reference data for the pictured idea...:" For those of us who have ever lived in a shared space know the sheer terror of the shared refrigerator. Not only do you go to your fridge to find that the food that you have been saving has miraculously vanished, but roommates tend to leave food in the fridge for so long, that even a forensics team doesn't want to touch it. Designer Stefan Buchberger used to live with a "dirty fridge in a shared flat", so he invented a device called the Flatshare that allows roommates to share a fridge while keeping their individual space. The Flatshare has a base station as well as four stackable modules that fit together like puzzle pieces. This makes the Flatshare very easy to transport when moving. The user has the option of putting colorful skins or a whiteboard on the modules. ....

130. Hexagon wood building stud

This idea designs a hexagon (6 sided) shaped building stud... This hexagon design buckles less under loads vertically and horizontally...It has more sides to attach interior building items....A typical 2" x 4" x 8' wood building stud wood be replaced with a 4" x 4" x 8' hexagon stud...Now shear walls can be reduced because the root core structure of the frame is reinforced with this hexagon stud.....There are approximately 750 million square acres of building wood in the United States so this resource is not an issue here....With this stud the contractor saves money on exterior sheathing shear walls...At 90 degree corners the contractor can use two hexagon's instead of the conventional three studs of 2" x 4".....

131. The singing hammer

This idea ergonomically designs a mechanism against the hammer head which is embedded inside the typical hammer... When the hammer head strikes an object there is a counter sound from this proposed mechanism which sounds nice...This is a white noise device to counter act the sound of a pounding hammer....

132. Mouthpiece nail holder

This idea ergonomically designs a mouthpiece, something like a sports mouthpiece, except this design protrudes out with a holder for nails...Let's face it a lot of guys hold nails in there mouths while they are hammering...Nobody produces an ergonomic mouthpiece to hold nails.....

133. Tool rental vending machines inside big box stores

This idea is to ergonomically offer the consumer tools for rent inside the big box stores via vending machines...An electronic chip monitoring system tracks the tool as well as an upfront credit card number in case its lost.....This idea benchmarks to the product below which offers bikes for rent...My idea houses the tool vending machines safely inside the big box stores....This idea also benchmarks to grocery stores offering DVD movie rentals via vending machines....I do believe all of us craftsman would jump at the chance of renting this tool and that tool from a company, even if it's for the pleasure of use for a little while and before we buy it......Respectfully....Reference data:...Getting around the city on a bicycle is a healthier and more environmentally friendly alternative, and the design of this urban Bicycle Rental Stand will surely help in generating a positive image. This design has already picked up an award at the 2007 Spark Design & Architecture Awards, working like a U-Haul that supplies bicycles in a one-way trip. Each bike comes equipped with RFID chips for easy tracking, enabling you to rent one at an affordable price. Once you're done, just return it at a nearby machine and you're good to go. Just make sure you take proper care of them and not return a punctured tire in the process.....

134. GREEN product cleaners & oils

This idea benchmarks to the Company "Green Earth Technologies" which offers various products that are environmentally friendly..These include motor oils and other cleaners....I feel a company should get on board with these products and offer a line of there own, or buy this company...This idea cannot go wrong because our planet needs as much help as possible in conserving resources....Please review there company profile.. .....Welcome to Green Earth Technologies, Inc (GET). Our "ecological" products now make it easier for you to GETGreen!™ and, together, we can stand proud knowing that we are making a difference. Our products are made entirely from American grown base oils that are enhanced with the power of nanotechnology and dehydrogenation. Our methods are patented and proprietary - the result is superior performing products that are TOTALLY GREEN! You will not have to sacrifice value or performance in order to GETGreen! and do your part......hence our motto: SAVE THE EARTH – SACRIFICE NOTHING™.....All our products are Ultimate Biodegradable (ASTM Standard's highest available ranking) which means they are environmentally safe....safe for you, safe for your family, safe around your pets and safe for the earth....GET's "G" Brands will help you in several areas of your life, offering products for your home, lawn, automobile or truck and boat, including multiple grade motor & engine oils, fuel additives, cleaners and solvents. We currently offer G-OIL™, G-DISPOSOIL™, G-WASH™, G-GLASS™, G-SCENT™, G-WHEEL™, G-TIRE™ and G-TILE™.... Please bookmark www.GETG.com and visit us frequently to see what's new. Our staff and alliances of Doctors, Scientists and

Professionals, are constantly doing research, development and testing of new products that work so well that it's easy being green...GETGreen !™.........

135. Hardware storage container with digital counter

This idea ergonomically designs a cheap hardware container with electronic digital counter....The user simply drops in there hardware and get an L.E.D. readout of that count..The L.E.D. can be turned on or off....The idea is great for neat freaks in there workshops....This idea benchmarks to the digital piggy bank pictured and noted below....Respectfully...Digital Counting Piggy Bank September 12th, 2008 by Edwin in Children's Gadgets... It is always good to get your kids started off on the right footing when it comes to money, and what better way to save than to watch their savings grow in due time? With the Digital Counting Piggy Bank, it makes saving all the more fun since you'll be able to know the exact amount inside without having to painstakingly count every single coin...... Piggy bank counts and keeps track of spare change for you! Not your typical "piggy" - electronic slot and LCD screen automatically track and display your total savings. Twist-off lid for bulk deposits and withdrawals. Features count-up/count-down controls to keep total accurate. Great for teaching kids to add and save money. Requires two AAA batteries (not included). ..... Since the thing is made out of plastic, you don't have to get a hammer to access your hidden stash during times of emergency thanks to the twist-off lid. You can pick up the Digital Counting Piggy Bank for $29.98 and get Junior on the saving program early in his life....

136. Construction site time-lapse camera cop

This idea is to ergonomically design a heavy duty time-lapse camera which can be mounted high against a typical large construction site backdrop....The idea is for the directors and city officials of the project to be able to review the work as it progresses...This also leaves evidence of proper construction methods....This product could be sold to cities when they are working on big construction projects...Usually inspectors go to the sites periodically to inspect and sign off...Now this camera also helps keep the contractors honest all the time because they are being pictured every few minutes......

137. Vehicle anti tip cargo net on passenger side

This idea ergonomically designs a cheap cargo net which is attached to the typical passenger side seat and hand safety handle atop the air bag at the dash....The user can now temporarily set items vertically so they do no tip over...For example, when i gave my girlfriends mom some flowers in a vase yesterday, she had to drive off with the flowers and vase between her legs so it would not tip over..Many other items cannot be tipped either...This anti tip cargo net would be very handy.....

138. Wood joinery system on dwelling construction

This idea is to use wood dowels in lieu of metal nails to join rough framing in the wood construction field....This idea is to design a drill to make dowel holes into all the aspects of rough framing...Another tool press fits the dowels for strong joints... Other tools do fast tongue and groove joints...The Stickley furniture company of New York proves wood joinery can last forever... Let's translate this idea to the construction field....This idea would be revolutionary for the Green movement in conserving this planets resources....Please read below for wood facts..... Respectfully.....Reference data:....."""Wood may be the most environmentally friendly material available for building homes or businesses. Here's why. ....Non-wood products are environmentally expensive. The supplies of ores and petroleum for their production are finite; once gone, they are gone forever. Wood, on the other hand, is a renewable resource. Non-wood products require far more energy to manufacture than wood: for example – nine times as much for a steel stud as for a wood stud. That further depletes supplies of fossil fuels and coal, not to mention increased pollution of the air and water... Wood is reusable, recyclable and biodegradable. Inorganic materials not only require excessive energy to produce, but also to recycle or dispose of when their use has been terminated.* .....A recent study from the Society of American Foresters reports that the United States has about 750 million acres of forestland, a number that has remained relatively stable for the past 100 years. Other positive notes include:.... On average, 11 percent of the world's forest-land benefits from some type of conservation effort;... Historical trends indicate that the standing inventory (volume of growing stock) of hardwood and softwood tree species in U.S. forests grew 49 percent between 1953 and 2006...... An estimated 25 percent of U.S. private forestland is managed in accordance with one of the three major forest certification programs ? Sustainable Forestry Initiative (SFI), Forest Stewardship Council (FSC) and American Tree Farm System .....

139. Pneumatic nail gun designed like G36 rifle

Current heavy duty nail guns are not comfortable...This idea is to design a heavy duty nail gun like the worlds most comfort-able assault rifle which is the Heckler and Koch G-36 ..I believe a new nail gun designed in this fashion will be a hit...I do believe the craftsman will love the ergonomics of the G36 design....If we chisel a few of the militaristic characteristics away we end up with a very comfortable pneumatic nail gun......Respectfully... Reference data:.... View Poll Results: M-16A2 5.56mm 18 8.57% M-4 5.56mm 29 13.81% SIG-552 5.56mm 9 4.29% G-3 7.62mm 16 7.62% G-36 5.56mm 47 22.38% AUG 5.56mm 6 2.86% FAMAS 5.56mm 4 1.90% FNC 5.56mm 3 1.43% L-85A2 5.56mm 7 3.33% IMI Galil 5.56mm 3 1.43% AK-47 7.62mm 41 19.52% AK-74 5.56mm 9 4.29% AK-101 5.56mm 3 1.43% AK-103 7.62mm 15 7.14% TKB-517 7.62mm 0 0% .....The G36 is a German 5.56 mm assault rifle, designed in the early 1990s by Heckler & Koch GmbH (HK) and accepted into service with the German Armed Forces in 1997, replacing the 7.62 mm G3 automatic rifle.[1].....

140. Miniature metal detector on wall stud sensors

This idea simply ergonomically adds a small metal detector to a stud sensors in order for the user to know if there is a metal connection somewhere along the vertical line of the stud...With all of today's add on one never knows what is behind the wall so to peak...The user sweeps up and down the wall and the metal detector sounds along the path so the user knows how big the piece of metal is...

141. Voice command opens lid or drawer on toolbox

This idea simply adds voice recognition to a miniature mechanism to open the lid or a draw on a toolbox...The toolbox can be a hand held or heavy duty...Your hand are full and you want to store something in your toolbox, just for it to open the lid or drawer.....The second version of this idea is for the user to ask the toolbox to open the drawer with the least weight...The system weighs each drawer and opens the one with the least weight, meaning the one with less items stored so the user can easily store new items...

142. Keyboard sensors determine workload productivity

This idea is not to enslave anybody but to ergonomically design a computer keyboard which detects actual time of user use.....The reason for this product would be to help human resources departments determine work load productivity and the best way of employing people who work on computers......

143. Site bicycle that is biodegradable & no flat tires

This idea ergonomically designs a special bicycle for the construction site...This bike is made of biodegradable materials, it has baskets and is weatherproof....The tires are run flat type from the automakers...I noticed on the bigger construction sites people using vehicles for short distances...Why not offer a green solution to moving small items..This bike could also be a tricycle with a larger holding basket....The materials for this bike could include hardwood dowels....Respectfully....Reference data:..."DETROIT, Mich.--(AUTOMOTIVE WIRE)--May 7, 1998-- Bridgestone/Firestone, which introduced its first run-flat tire in the U.S. in 1992, announced today that this summer it will launch the Firehawk SH30 RFT, a tire designed to travel with no air pressure for up to 50 miles at 55 mph using conventional wheels.

The H speed-rated Firehawk SH30 RFT includes Bridgestone/ Firestone's exclusive UNI-T(R) technology and will initially be available in three sizes designed to fit a wide variety of today's vehicles. The product will be sold through Bridgestone/Firestone certified run-flat retailers.....

144. Panic button turns off tool and sounds alarm

This idea ergonomically designs a red panic button like the one pictured in attachment...This button is located conveniently for easy access on heavy duty power tools....When this button is punched the tool shuts down and alarm begins to sound..... Story, we once had a manufacturing engineer cut his finger off on a pallet Jack, when people finally saw him he was doing circles bleeding profusely...If an alarm would have sounded then more attention would have been paid by others nearby.....

145. "Green wood nail" for nailing guns & standard

This idea designs a typical wood nail except the material for this nail is hardwood...This nail has a steel tip only for dynamic entry into objects....This nail is meant for pneumatic nailing guns but could be sold separately for standard hammering....I believe a hardwood can be found which is strong enough to be shot by a nailing gun and to endure as an actual nail.....This idea would be revolutionary for the Green movement in conserving this planets resources....Please read below for wood facts..... Respectfully.....Reference data:...."""Wood may be the most environmentally friendly material available for building homes or businesses. Here's why. ....Non-wood products are environ-mentally expensive. The supplies of ores and petroleum for their production are finite; once gone, they are gone forever. Wood, on the other hand, is a renewable resource. Non-wood products require far more energy to manufacture than wood: for example – nine times as much for a steel stud as for a wood stud. That further depletes supplies of fossil fuels and coal, not to mention increased pollution of the air and water... Wood is reusable, re-cyclable and biodegradable. Inorganic materials not only require excessive energy to produce, but also to recycle or dispose of when their use has been terminated.* .....A recent study from the Society of American Foresters reports that the United States has about 750 million acres of forestland, a number that has remained relatively stable for the past 100 years. Other positive notes include:.... On average, 11 percent of the world's forestland benefits from some type of conservation effort;... Historical trends indicate that the standing inventory (volume of growing stock) of hardwood and softwood tree species in U.S. forests grew 49 percent between 1953 and 2006...... An estimated 25 percent of U.S. private forestland is managed in accordance with one of the three major forest certification programs ? Sustainable Forestry Initiative (SFI), Forest Stewardship Council (FSC) and American Tree Farm System (ATFS). ....

146. Soil & water tester at big box stores

This idea to develop a machine that can test for soil and water pollutants...This machine could be sold to the big box stores so the consumers cold bring in a sample and have it tested locally...Right now if I buy a lot or want to check the soil or water around my home I really don't know what to do?.... Why not start to ball rolling in this area?...At first it does not have to be very a elaborate system/machine but at least it shows progress in a GREEN corporation,.....Respectfully..... Reference data:...."Advanced Water & Soil Analysis... Torrent is a California State Certified Analytical Laboratory with advanced sampling, testing and reporting for water and soil analysis. Our chemists perform Volatile Organic Compounds analysis with next-generation Gas Chromatography and Mass Spectometry systems. We offer Northern California's most highly automated extraction capabilities in laboratory analysis for Semi-Volatile Organics Compounds, and the region's most complete laboratory services for complex testing and analysis of Inorganics samples. VOLATILE ORGANICS SEMI-VOLATILE ORGANICS INORGANICS TPH gas range organics 8270 - 8270 SIM Metals by simultaneous ICP Carbon speciation PAH (PNAs) - also by SIM Metals by ICP MS Fixed Gases GPC Clean-up Perchlorates RSK-175 Pesticides Hexavalent Chromium Non-Methane Organics PCBs Total Organic Carbon (TOC - Soil/Water) 8260 Water and Soil Floricil Clean-up

Cyanide 8021B Diesel Motor Oil / other fuels Oil and Grease 8015 -> TPH Purgables Heavy Hydrocarbon Fingerprinting California WET STLC Special Methods Development Silica Gel Clean-up TCLP Water Quality Parameters Custom method development STORM WATER ANALYSIS Maintain compliance with state storm water monitoring regulations.....

147. Pliers w/independent teething mechanism

This idea designs a plier except each tooth runs on independent racks and mold themselves to the object they are gripping...I believe with each tooth adjusting to the object this tool will have more gripping power......Respectfully....Reference data:..." 84-881 - 6" MaxGrip™ Pliers...Print Printer Friendly Version Email to a friend Displaying product 1 of 3...images/NextPageIcon.gif Click to Enlarge Click to Enlarge Features and Benefits * Self adjusting 6" Pliers * One hand adjustment up to 1-7/16" * Dual material handle for superior gripping * Ergonomic shape for maximum comfort in all applications * Handle hole for tethering or hanging * Patented design ......

148. The copper tubing handle utility knife

This idea designs a typical utility knife with the handle being copper tubing...The copper is flexible and so would allow some degree of curvature needed for certain specific cutting scenarios...The is an ergonomic design which would look nice......

149. Motor helps "C" clamp threads move in and out fast

This idea ergonomically adds a mini motor and on/off button to turn the threads in and out fast....The clamp is still hand tightened...The reason is to save time in turning threads...This idea can transcend to your other clamps...Since I can remember I hate turning the clamps, seems like time is wasted, let's make it faster....

150. Proximity sensor alarm worn on body & product

This idea ergonomically designs a proximity sensor alarm system users can attach to items such as portable laptop computers and on there person.....When the user is a certain distance away from the item being protected the sensor sounds to warn the user they left that item behind...The sensors are attached to the user and the item being protected....To many of us leave items behind....Why not offer a tiny yet powerful tool to give us the heads up we have left it behind?...Proximity sensor distance in feet can be graduated...This item is great for school and traveling.....

151. Hand operated drain unclogger - A Green tooL

This idea designs a plumbing drain cleaner / unclogger which is operated by a hand crank mechanism, similar to a hand drill... The end of the shaft on this tool has screw threads to help unclog even better....This tool is 18 inches long....All the materials on this tool are biodegradable...This idea benchmarks to the company "Super tool" and there "drainstick"....Respectfully...Reference data"..."" DRAIN CLOGS. Using toxic chemicals to clear a clogged drain is anything but green. The alternative is to call a plumber or attempt to clear the clog using a drain snake, which can be cumbersome at best and lead to a real mess..... The folks at Superior Tool have come up with an alternative called the Yellow Submarine Power DrainStick: a thin 18" flexible plastic snake that is connected to a mini screw gun that fits neatly in the palm of your hand. Simply insert the snake into the drain and pull the trigger. No chemical, no plumbing bills..... Superior Tool Company #03819 Drain Stick SKU: sku-0074677-00 Availability: Usually ships in 1-2 business days... Drain Stick, Pushes Through Sink, Bath & Shower Drain Clogs, Tough Nylon Construction Will Not Rust, Reaches Where Needed, Easy To Use With No Necessary Disassembly Of Pipes, No Mess, Dangerous Chemicals Or Heavy Snakes/Augers, Rinses Clean, Made In The USA....

152. Hand lawn mower for aerating - Organic Green

This idea designs a hand lawn mower except with spikes attached instead of cutting blades...These spikes poke holes into the lawn as the user....The reason for spikes is for aerating the typical grass lawn...This idea comes from the new Green movement to help preserve the earths resources.....Please read below for all the benefits of aerating lawns....Respectfully...""""
Lawn aeration is the secret weapon in making your lawn healthy. Aeration involves making holes in the lawn either by pushing a rod into it or by "coring", extracting a plug of soil. By aerating your lawn you provide the following benefits to your lawn and its root system: * Oxygen gets to the roots and the soil allowing it to "breathe" * Organic fertilizers and nutrients get access to the root system * Water is able to better soak the soil and reach the root system * Helps to break up thatch * Tight, compacted soil is loosened up allowing the root system to grow...... * Be careful of shallow sprinkler lines getting punctured by the aerator. * Water your lawn heavily, up to an inch, the day before you plan on aerating. The soil should be soft and moist when aerating but not muddy. * Aerate in the Spring or late Summer or Fall. * In arid / dry climates aerate twice a year. * Don't waste your time or money with shoe aerators. * Whenever possible, use a coring aerator as opposed to a spike aerator. Spike aerators just further compact the soil while making a hole. * After core aeration, leave the plugs on the grass and allow them to dry out, then rake the plugs into the grass if you like. It's OK if you don't as they will break up after the first mowing and help to breakdown thatch by providing micro organisms that will feed on thatch. * Be careful of shallow sprinkler lines getting punctured by the aerator. * Water your lawn heavily, up to an inch, the day before you plan on aerating. The soil should be soft and moist when aerating but not muddy. * Aerate in the Spring or late Summer or Fall. * In arid / dry climates aerate twice a year. * Don't waste your time or money with shoe aerators. * Whenever possible, use a coring aerator as opposed to a spike aerator. Spike aerators just further compact the soil while making a hole. * After core aeration, leave the plugs on the grass and allow them to dry out, then rake the plugs into the grass if you like. It's OK if you don't as they will

break up after the first mowing and help to breakdown thatch by providing micro organisms that will feed on thatch.....

153. Cylindrical plumbing tube to prevent marring

This idea designs a cylindrical tube which is hollow and has a slit for adjusting the diameter around a given plumbing tool... The purpose of this design is to prevent marring plumbing objects when they are worked on.....The material of this tube to match the properties of duct tape except stronger as not to decay.....Respectfully....Reference data:..."""""When using pliers to remove a slip nut on a sink waste trap, removing a shower head, removing a tub spout or any other item where you want to prevent marring of the finish, duct tape can be used to cover the jaws of the pliers and protect the finish, or wrapped around the item itself.... * When using duct tape to line the jaws of a set of pliers, cut small strips of duct tape to the appropriate length and width of the pliers * Use about 3 to 4 layers of tape on each jaw if highly textured like these pliers * Use 1 to 2 layers if there are no texture or light texture on the pliers jaws.....

154. Manually adjustable flathead screwdriver tip width

This idea ergonomically designs a flathead screwdriver whose tip is adjustable to different widths....This screwdriver has a thumb wheel at the top near the handle, similar to the ones on crescent wrenches....Now the consumer does not have to fiddle around finding different sized flathead screwdrivers...If this does not make sense I can make a sketch......Respectfully......My idea came from this quote from a woodworking magazine:..."""""When using a flathead screw, choose a screwdriver whose tip is not wider than the screw, as the screwdriver will end up marring the wood as the screw is set. If the head is too narrow, the screw may strip. Woodworking Joinery....

155. The hammering robot

This idea designs an ergonomic hammering robot for field use...This robot hammers all types of nails where indicated by user via a special marker...For example, drywall needs to be hammered down in a garage...The workman would position the drywall square with a couple of nails only then mark the places to be hammered with the special marker...This robot would then go about hammering down the drywall at the places indicated....It has multi magazine clips with all different types of nails.....I has wheels and is mobile...It has an extended shooting barrel which can extend many feet to reach two stories and deep into crevices....This type of robot would help out tremendously and keep people safe...It runs on pneumatics and solar/ battery power...It is rechargeable...Revolutionary for the industry.....

156. Dual folding tape measure

This idea designs two tape measures side by side with an ergonomic hinge between them...With this new design the user can measure to 180 degree positions or two items side by side as sketch.....Novelty or functional, let the consumers decide...

157. Miniature cooling units for power tools

This idea is to let you know engineers are working on miniature cooling units that act like little refrigerators...If you can benchmark to this study and develop mini-cooling units for power tools to dissipate heat when running and cutting.........Reference data:...." Mechanical engineers at Purdue University have new findings offering promise for modifying household refrigeration technology with small devices to cool future weapons systems and computer chips. The devices, called "micro-channel heat sinks," circulate coolant through numerous channels about three times the width of a human hair. Such devices might be attached directly to electronic components in military lasers, microwave radar and weapons systems, as well as in future computers that will generate more heat than present computers, said Issam Mudawar, a professor of mechanical engineering who is leading the research....The researchers are adapting refrigeration systems by using the micro-channel heat sinks to replace conventional "evaporators" – components in household refrigerators that contain a labyrinth of tubing. As coolant circulates through the tubing, heat is removed from the refrigerator to cool the food inside. "We are substituting these conventional evaporators – which might be well over a meter long in the typical refrigerator – with a heat sink that's only about 1 inch square," Mudawar said. "The challenge is how to unplug this large evaporator and put in its place this tiny heat sink and make the whole system work."... Recent findings were detailed in two research papers that appeared in the February issue of the International Journal of Heat and Mass Transfer. The papers were written by mechanical engineering doctoral student Jaeseon Lee and Mudawar. Electronics for new weapons systems, as well as chips in future computers, will generate five to 10 times more heat than chips in conventional electronic products, requiring better cooling systems. Computers and other electronic equipment are typically cooled with bulky assemblies that use metal fins to dissipate heat and fans to circulate the hot air away from components. But electronic components in new weapons systems, such as advanced lasers and chips in future computers, will generate too much heat to be cooled with conventional systems that use fans, Mudawar said....One possible solution is a "two-phase" cooling system – the same basic technology used in a conventional refrigerator – in which a liquid coolant absorbs heat, turns into a vapor and is then pressurized by a compressor and condensed back into a liquid to begin the cycle over again....

158. Tape measure with round tape reel

This idea designs a tape measure except the pull out reel is round and made of the same metallic material as the conventional ones with the capability of easy bending...This design looks like a tape measure with a thick string coming out the end....Incremental measurement markings on the reel at 360 degrees......I believe this to be an ergonomic idea to view measures....If this is not clear I can make a sketch.....

159. Toolbox camouflage as living space furniture

This idea ergonomically designs a toolbox which is camouflaged as a piece of furniture...The attached picture is the perfect example to benchmark....Now the craftsmen who have been running out of room can beautifully hide there tools inside a living space without the visitors knowing what they are storing...The folks with small houses or condo's can do the same...The exact piece of furniture to be determined.....

160. Electric tools operate with battery pack detached

This idea is to detach the battery pack from cordless tools for less weight handling by the consumer yet still able to run that tool......The battery still sends electrical power to the given tool through an electrical wire which winds in and out as the user walks around and performs there task...For example, i have to perform a few task with my cordless tool in a room...I detach the battery from it and leave it in the middle of the room, now the cord winds and unwinds while i do my tasks...The user deals with less weight which is still a factor with cordless tools...

161. Work boots with heel storage

This idea ergonomically designs construction work boots with heel storage...The heels are removable to hide and store small items such as keys, drivers license and credit cards......Please note the construction site is a rugged place...We don't like loosing our wallets there...

162. Remote control opening drawers in toolbox

This idea adds remote control capability to open any or all the drawers on a heavy duty tools box...When you hands are full of tools wouldn't it be nice to open the drawers on your toolbox remotely?...I think so...The proposed opener to be ergonomic and miniature....This device could be sold separately and or installed in a dedicated heavy duty tool box....Respectfully..... Reference data:....."""You attach the swinging door opener to the top of your door (and wall). You are then able to open the door via remote control. This could be useful for disabled dwellings or just plain OTT in most houses. Still looks like a cool gadget to me though :)...Available from SmartHome but be warned it's quite pricey. Found via Gizmos for Geeks.....

163. Rolling rulers and levels

This idea designs a rolling ruler and level...These two be round in shape and able to roll by hand.....Why push as done conventionally?....Now the craftsman has a three dimensional effect with his hands and vision with these new rolling tools...I believe this to be a crafty idea to give the user a more artistic and ergonomic way of taking measures.....Respectfully.....My idea benchmarks to the rolling pin idea as reference:...."Ingeniously simple, the Rolling Pin Ruler. Next time your baking and the recipe says roll out 15inch squares of pastry, instead of just guessing you can measure it with the rolling pin. Can be used to measure up to 12 inches (30cm). It's a good job it's labeled rolling pin or visitors might mistake for something else (though I've never seen one with measurements on the side). Available from Drink Stuff as drinking and burning baking are so closely connected.....

164. Dummy tool as a safe to store valuables

This idea designs a tool which is actually a safe...The geometry of the safe is the tool...In other words a heavy duty drill actually opens from the bottom so the users could store valuables....The walls could be rated fireproof as well....I note when I sold Stickley furniture they had pieces like chinas and chairs with hidden compartments...No thief wold think a tool is a safe....A hammer, a drill even a tap measure could be the dummy tool..... My idea benchmarks to the campbell soup can as referenced below........Reference data:..."At first glance, this can of Campbells soup might look pretty ordinary, but it is actually a safe that lets you store your petty jewelry without making it look like a prime suspect during a burglary. This will probably be the last place the thief will want to check, unless he is famished and has a love for canned soup. You ought to be able to store a fair amount of dough inside in rolls, so the £12.49 investment is well worth the price in terms of safety. You can also choose from the WD-40 or Sprite can safes......

165. Flashlight with integrated video recorder

This idea designs a heavy duty waterproof flashlight with integrated video recorder....Now the user has dual functions with there flashlights...The uses are many plus the novelty....This idea benchmarks to the Japanese Company and there version as referenced below.......Respectfully.......Reference data:..."""If only the old pulp novelists had gadgets like this to work with. Can you imagine how much shorter mystery stories of the past would be with modern forensics? Forget the heavy thinking detective. Technology is the Sherlock Holmes of modern times....Japanese company Carrot Systems (apparently ran by bunny rabbits) has produced a flashlight with an integrated video camera and a sleek streamline design. The flashlight is touted as an awesome tool for crime scene investigators who want to capture a dynamic peek of any dark and seedy crime locations......

166. Tumbler storage bins for misc hardware

This idea designs storage bins designed around a central horizontal beam tube..I call these bins "tumblers".....Each tumbler can rotate 360 degrees by the human hand...Each side of the storage tumbler has a door and compartment.....Now the user can store more misc components in less space....Just a curious design....There is almost no limit to the number of tumblers that can be set side by side...Maybe these tumblers could be sold separately for the user to assemble as many as they wish....This design can also be stacked in horizontal rows....

167. Mark correct hand position on hammer handles

This idea is to mark on the typical hammer handles the correct place to grip the handle for maximum performance (which is near the end)...I realized yesterday many people do no know this fact...My calculation states hammer blows can go to 300 pounds of force from the end for max performance...Let's educate the users....Attached sketch is a tracing of my fatmax 24 hammer handle. I propose a marking near the end as shown...In different languages too......

168. Work belt to hold heavy nailing guns - hip shots

This idea designs a special tool belt with a side to side adjusting ergonomically designed pouch...This pouch is designed to hold the butt of the typical heavy duty nailing gun....The framing nailer gun weighs approx. 7.9 pounds...This new belt allows the users to take hip shots with there nailing guns once they get tired of holding it conventionally...The pouch adjusts on the tool belt by hand to fit the users personal body frame.....This belt and pouch could be made out of leather....

169. The mechanical foot wedge

This idea designs a foot operated mechanism to use as a wedge against items needing a small lift....The user would place one end under the item to be tilted and with the other foot presses down....Now the housekeepers can vacuum and clean under big objects.....

170. The wing nut socket ratchet

This idea designs a new ratchet that looks like a wing nut... It's mechanism lets the user screw and unscrew bolts...This tiny tool is operated with the human fingers...These come in sets for different sizes...Sometimes we need tiny tools to bolt and unbolt...

171. Tape measure with double reel measurements

This idea is a tape measure which has two tape reels, each going in opposite directions...The user can select for just the left reel out or right reel out or both reels out...This is great for measuring interior dimensions of an object such as a drawers.....The readout is digital...It has a three position switch atop to control the reels and explained. This tape battery operated for the reel or hand operated, at users discretion...

172. Eleven Bravo heavy duty Lunchbox

This idea is a super heavy duty lunch box for the construction workers...This lunch box has everything needed for a meal in the field...This box is portable and has a fixed handle and a folding handle...It also has big rubber wheels...Please remember this box is not meant to be pretty but rugged...It has no windows...I has hasps not door hinges that can fly open.....I has a recharge-able battery to run all the functions...On top it has a flip up door which reveals utensils and a cutting board....A hand sanitizer is integrated into this unit...It includes a radio, refrigerator, toaster, coffee maker and microwave oven....This unit is not large but a perfect size to serve two people comfortably...This unit can also be sold as an earthquake or natural disaster survival tool...

173. Thick paste wood glue with geometric tip bottle

This idea is a thicker paste carpenters wood glue with various geometric tips and bottle...With a thicker paste wood glue and special tip the carpenters can deliver the glue more precisely and with less spill over...This idea benchmarks to cake decorating kits with all the different tip configurations to deliver precise decora-tions...This idea also benchmarks to silicone tile grout applica-tors and caulk guns used in bathrooms and kitchens with a thick paste silicone....Since childhood i hated the way carpenters glues run and spill--- too fast...Let's improve this......

174. Add ergonomic handles to squares

This is to add ergonomic handles to your measuring squares...I note the current designs are old and could use some upgrading, handles on tools like these squares is a great start..My friends and I confirm a handle would be nice, no fumbling around with these low profile tools.......Respectfully....Reference:..46-050 - Adjustable Quick Square® Layout Tool....

175. Thermos case for hammers with wood handles

This idea ergonomically designs a thermos case to house the wood handles on the hammers. The reason for this product is to preserve the wood on the hammer handle...Uneven temperatures cause hammer handles to swell the wood fibers inside the handles head, crushing them, loosening the head and making handle replacement necessary...Extreme dryness, as on a shelf above a heater can shrink the handle which also causes looseness...The idea is to maintain even temps and therefore preserving the hammer handle forever.....Now the expensive wood hammer case a case to preserve them....Respectfully....Reference data:...""This thermInner Workings of a Thermos One way to build a thermos-like container would be to take a jar and wrap it in, for example, foam insulation. Insulation works by two principles. First, the plastic in the foam is not a very good heat conductor. Second, the air trapped in the foam is an even worse heat conductor. So conduction has been reduced. Because the air is broken into tiny bubbles, the other thing foam insulation does is largely eliminate convection inside the foam. Heat transfer through foam is therefore pretty small. It turns out that there is an even better insulator than foam: a vacuum. A vacuum is a lack of atoms. A "perfect vacuum" contains zero atoms. It is nearly impossible to create a perfect vacuum, but you can get close. Without atoms you eliminate conduction and convection completely. What you find in a thermos is a glass envelope holding a vacuum. Inside a thermos is glass, and around the glass is a vacuum. The glass envelope is fragile, so it is encased in a plastic or metal case. In many thermoses you can actually unscrew and remove this glass envelope. A thermos then goes one step further. The glass is silvered (like a mirror) to reduce infrared radiation. The combination of a vacuum and the silvering greatly reduces heat transfer by convection, conduction and radiation. So why do hot things in a thermos ever cool down? You can see in the figure two paths for heat transfer. The big one is the cap. The other one is the glass, which provides a conduction path at the top of the flask where the inner and outer walls meet. Although heat transfer through these paths is small, it is not zero. Does the thermos know whether the fluid inside it is hot or cold? No. All the thermos is doing is limiting heat transfer through the walls of the thermos. That lets the fluid inside the thermos keep its temperature nearly constant for a long period of time (whether the temperature is hot or cold). os sheath would keep the temperature even and so preserve the hammer forever....

176. Drill bit stand with flute wear detector

This idea ergonomically designs a drill bit stand which informs the user when the drill flute needs sharpening...This stand has silhouette disks at the bottom of the storage cylinder for each drill bit size....Once the drill bit is placed inside its cylinder storage in the stand, then the laser detector compares the silhouette to the actual drill flute, thereby giving a reading of flue wear....The results information can be with L.E.D. light reading:... "GOOD"... "WORN"..."NEEDS SHARPENING".....

177. Anti-marring plate sold with wood hammers

This idea ergonomically designs anti marring plates for wood hammers...This plate has a small handle to help the user place this plate under the hammer head so it won't mark the wood when the claw pulls the nail out of the wood....Different sizes to go with the different wood hammer sizes.....I picture a small key chain on the hammer with this plate attached in the stores.....

178. Wood nail with loctite glue pouch under nail head

This idea adds a tiny circular pouch right under the typical nail head with the same radius as the nail head, material can be breakable plastic.....The idea is for the user to nail down there wood nails as usual...Once this nails head hits the surface of the wood it's loctite glue pouch breaks open and in so glues the steel nail head to the wood, an added gripping factor....This tiny pouch under the nail head is not a hindrance to nailing.....Now we have a nail dedicated to extra gripping power....Exact loctite glue to be determined.....

179. Dual sided pen to write on wood and steel

This idea is to have a heavy duty writing pen with two tips at opposite ends...One end is for writing on steel and the other end for writing on wood....Something like a sharpie for the wood writing..For the steel writing a new ingredient would have to be developed, a cross between correction fluid, paint pen and sharpie..We all need this new pen!!.....Respectfully..... Background on this idea"...What do you guys use to write, or draw on steel? Something that works on a clean slate, and also on something rusted, or coated in that new metal oil? So far I use felt pens, and some silver pencils that work alright. But nothing that works well on both.""....have tried the soapstone, but it wears away too easy...""""i have heard the silver paint pens are the cat's ass, i need to try them.I often use a pic and scratch a line instead of drawing one."..."To just write something down... we use paint pens mainly... We'll use sharpies on sheetmetal work and some tube work. Red shows up real well".....""silver colored pencils work realy well, ultra fine sharpie kick a$$ for sheetmetal. for nasty metal regular sharpie work wonders and for huge hunks of junk that you are just writing info on white grease pens are excellent".....""metal scribes and dividers are the prefered method where i work. we have a 1/64th inch tolerance so that's about as fine a line as you can get.""...."I use the silver/grey sharpies - I like them best, nothing else seems to hold up. They even work when torch cutting - the ink stays under the heat.""....We've tried TONS of different markers and pens. The silver sharpies didn't last for shit for us. Regular sharpies have a pretty good life but are hard to see on dark metals. One tip if you are using a sharpie, don't throw it out when the tip gets dry or full of crap and won't write anymore. Instead lightly touch it off with a flap wheel on a grinder. Takes away the crappy felt and it should work like new again. As of right now we use white paint markers. They work great and it takes more to wear the line off which has been a problem before with the sharpie. I have a bunch of standard ones that we use now and like. At Fabtech there were a couple of booths selling markers that are supposedly designed to write on steel. I picked up some free samples, black and yellow, Brite-Mark made by Dykem. So far I really like the yellow, but they have a strong xcelen smell. I'll probably just go get some cheapo yellow paint pens when I use up the overstock I have right now, works the same anyway.....

180. Laser pointer digital protractor

This idea is to offer a laser pointer to a digital protractor...I can see many uses for this unique tool......I found this comment as reference".....".But with maybe a magnetic base, then with a swivel head that has a laser pointer...Point being, you could find what angles to notch any tube, in any place, in less than a second...Basically put the device where the tube would start, then point the laser across the car to where the other side would be, and it tells you the angle?...."

181. Security motion sensing light with laser pointer

This idea adds a laser pointer to a motion sensing security light...The laser pointer would simply follow the source of the alert , then follow this alert target while being tracked by the motion sensor...This gives a very visual display of where the target of intrusion is located....Even a police helicopter for example would see this laser....The laser pointer is activated to work with the sensing light at users will....I believe this adds a nice twist to the security light.....Respectfully...Reference data:...""""The "motion sensing" feature on most lights (and security systems) is a passive system that detects infrared energy. These sensors are therefore known as PIR (passive infrared) detectors or pyroelectric sensors. In order to make a sensor that can detect a human being, you need to make the sensor sensitive to the temperature of a human body. Humans, having a skin temperature of about 93 degrees F, radiate infrared energy with a wavelength between 9 and 10 micrometers. Therefore, the sensors are typically sensitive in the range of 8 to 12 micrometers. The devices themselves are simple electronic components not unlike a photosensor. The infrared light bumps electrons off a substrate, and these electrons can be detected and amplified into a signal. You have probably noticed that your light is sensitive to motion, but not to a person who is standing still. That's because the electronics package attached to the sensor is looking for a fairly rapid change in the amount of infrared energy it is seeing. When a person walks by, the amount of infrared energy in the field of view changes rapidly and is easily detected. You do not want the sensor detecting slower changes, like the sidewalk cooling off at night. Your motion sensing light has a wide field of view because of the lens covering the sensor. Infrared energy is a form of light, so you can focus and bend it with plastic lenses. But it's not like there is a 2-D array of sensors in there. There is a single (or sometimes two) sensors inside looking for changes in infrared energy. If you have a burglar alarm with motion sensors, you may have noticed that the motion sensors cannot "see" you when you are outside looking through a window. That is because glass is not very transparent to infrared energy. This, by the way, is the basis of a greenhouse. Light passes through the glass into the greenhouse and heats things up inside the greenhouse. The glass is then opaque to the infrared energy these heated things are emitting, so the heat is trapped inside the greenhouse. It makes sense that a motion detector sensitive to infrared energy cannot see through glass windows....

182. Tape measure with giant digital readouts

This idea designs a tape measure with a giant digital readout.... This idea just sweeps across the top with bi numbers indicating tape measure readout....No tape recorder on this idea... Respectfully......Reference date:..."""Bob Vila 25' Digital Tape Measure with Voice Recorder Item: 183-576...Get the dimensions of your next home improvement project down pat with this Digital Tape Measure by Bob Vila. It combines a rugged and practical design with intuitive technology that makes it easy to size up your work. The LCD measurement readout display means you don't have to calculate fractional units of measure. Your tape measure is engineered so you can ready both the measurement on the tape and digital readout right-side up. With just a little practice, you'll know by touch which buttons to push as you measure for your projects and record notes to yourself. It can record two 5-second messages, so you can take the device with you when you go to get your lumber, purchase your curtain rods, etc.

Bob Vila Digital Tape Measure Features: * Built-in voice recorder - lets you record up to 2 callouts and stores them in memory for later playback * Triple rivet hook with blade protector - delivers improved strength and durability * Integrated shock absorber - protects the blade tip from damage * Nylon-coated blade - provides improved resistance to rust and corrosion * 25' high quality nylon tape * Full body rubber protective boot * Powered by 2 CR2032 batteries (included) * Steel, rubber and ABS construction * Black, blue and gray finish * Measures approx. 4-3/4"L x 3-3/4"W x 2"H * Comes with a manufacturer's 2-year limited warranty For warranty information, please call HSN.com Customer Service at 800.933.2887 (8 am-1 am ET)......

183. Water heater lift hand truck

This idea ergonomically designs a dedicated hand truck for lifting, moving and installing water heaters normally requiring two people to install....This hand truck would have two hand ratchets gearing systems, one for lifting a platform vertically about 30 inches or so..The other for moving the water heater horizontally forward Z9install position)....Now the user has a sort of vertical and horizontal elevator for moving and installing typical water heaters.....This tool is meant to be used by only one user....I want to add this tool would come in handy for many more items.....

184. Benchmark to brushless motors in power tools

This idea is for you to change up motor types on power tools.... It looks like brushless is the way to go and the one to advertise for increased sales to the heavy duty users........I found this info as reference.." I have always wondered why tool manufactures still keep using the same brushed motors. The brushless motor revolution is taking the radio control world by storm. Airplanes are flying over 200 mph, and electric cars are approaching 100 mph on race ovals. I know brushless motors cost more, but they would never wear out. Brushless motors combined with the newest batteries out(Li-Poly) would allow your cordless tool to last a long time on a single charge because brushless motors are over

90% efficient. I believe manufactures are somewhat holding back......."The fest tool company has brushless motors in some of there cordless drills. so i guess size is not an issue.....

185. Nail guns with auto sound when magazine low

This idea adds a sensor and horn on nailer magazines to tell the user when the nail count is low in the magazines.....An ergonomic subtle horn sound will tell the user that a reload is needed soon thereby giving the user time to plan the next move in there work and not over strain by setting something up with low nails in the gun......Reference data on a framing nailer......" F21PL2 - 21° Plastic Collated Framing Nailer Print Printer Friendly Version Email to a friend Displaying product images/PreviousPageIcon. gif 3 of 12 images/NextPageIcon.gif Click to Enlarge Click to Enlarge Features and Benefits * Patented push button adjustable depth guide sets nails to desired depths quickly & conveniently * Lightweight magnesium design for added durability * 16" layout indicator on magazine helps achieve correct distance between studs in a matter of seconds * Best in class power-to-weight ration (1,050 in/lbs) * Recommended tool for use with HurriQuake disaster-resistant nails Product Details Weight 8.1 lbs, 3.67 kilos Driving Power 1050 inch/lbs. Fastener Gauge .113" - .162" diameter Fastener Range 2" - 3-1/2" 21 degree plastic collated framing nails Magazine Capacity 60 Thread Size 1/4" NPT Application * Framing * Sheathing * Sub-Flooring * Metal Connectors * Bracing......

186. Nail straight pop back tin

This idea designs a flexible plate used to straighten out bent nails via hammer blows....This design bounces back after each hammer blow on the nail....The bent nail is set inside the pocket which will eventually straighten it...The user holds the handle and begins to hammer the nail to a straight position....Might work on nails under a 90 degree bend condition...The design is magnetized to further keep the bent nail inside the pocket.... Pocket sizes can vary with different tins, small, medium and large to cover the different penny nails.....The key here is the flexible bend back material used for this design......

187. The spiral tree tool holder

This idea designs a tool holder that spirals around any given tool.....I call it a tree because is has a stem on it that attaches to an agronomical base.....This stem is a bit flexible and so would bend when a tool is stored in it giving an agronomical zest to the look...This base can have locking wheels for easy movement around the shop then secured....The heights of the stem and the spirals can very...You can make a large design for pneumatic tools and a small design for small hand tools.....Mix match is the key here....The user can buy the base then buy as many spiral trees as they wish...Any tool fits nicely around this tree...Spiral diameters can vary, small, medium and large...I believe this to be a nice tool toy with functionality...

188. The "Von Resturff effect" for new hammer design

This idea designs a new hammer which integrates the following:. nail straightener, plier, reverse nail claw and a second hammer head on the back....This hammer design is fully loaded, any one of these integrated designs can be removed at your will...For example you might like the plier integration but not the others....This idea explains to you the "Von Resturff" effect which is a phenomenon of memory in which noticeably different things are more likely to be remembered than common things...

189. Digital time clock with words not numbers

This idea designs a digital clock displaying words not numbers...I believe a product this different could be a hit...A lot of people in "soft" situations might want to see a visual words display saying "nine thirty one" instead of 9:31....Respectfully... This idea benchmarks to my previous idea ..... Tumbler lock with pictures instead of numbers.....This idea is to design agronomical tumbler locks for the construction industry using pictures instead of number for the tumblers...It looks like Wordlock has a patent with there letter tumblers so I thought to go one further...

Actually humans think in pictures and so would be easier to remember than words....The exact pictures on the tumblers to be determined via psychology analysis.....Respectfully..... Reference data:..."Wordlock® Never Forget Your Combination The Wordlock® family of combination locks are the first combination locks that allows consumers to set their combination using words NOT numbers. These products are the next generation in personal security. Words are easy to remember, numbers are easy to forget and Wordlock® products are unforgettable. spacer All Wordlock® products are designed with the consumer in mind. Each product allows you to reset the combination to a word or a phrase that has meaning to you and is easier to remember. Resetting the combination is so simple you can change it anytime you want and as often as you like. spacer Easy to Set • Never Forget™ Wordlock® was chosen by the US Patent office and QVC as one of the top new inventions of the year. These award winning products have been shown on Regis and Kelly, Home and Garden TV, ABC news. ..

190. The new mouse trap

This idea ergonomically designs a new mouse trap...This design carries two mouse traps on one base and facing each other...This new trap has a handle and is made from a another material than wood...This trap is cleanable and made to be used over and over again...Experiments have shown that two mouse traps facing each other will have a better chance of catching a mouse...Another fact is that mouse's go where other mouse's have been, by smell....In other words once this proposed mouse trap catches a mouse, the consumer would dump that mouse and replace the trap to catch the others.....

191. Paper mache' concept for new "GREEN TOOL" line

This idea invents a new tool line to HELP the Green movement....This idea uses the paper mache' methodology to build some basic tools which are completely 100 percent biodegradable...If the material can be made hard enough then even screw drivers can fall into this new line.....An across the board tool study has to be made....Even just a few of the basic tools would suffice to start...Imagine a utility knife where only the blade is metal, the rest 100 percent biodegradable... This idea benchmarks to a previous idea of mine but did not offer the paper mache' concept.....

192. Steamer / Sanitizer for work tools

This idea is to offer an economical steamer /sanitizer for the work tool and construction industry....This idea benchmarks to the MONSTER1200 pictured except this version is more heavy duty type....Now all of us craftsmen in the United States can stay clean and sanitized with our tools....Please read below for this revolutionary tool.....Respectfully....Reference data:....." Sanitize Your Entire Bathroom in 15 Seconds * Removes Mold & Mildew * Kills 99.99% of Germs & Bacteria * Cleans & Deodorizes * Deep Clean Any Surface from the Kitchen to the Bathroom Product Includes: * Monster 1200W Cleaning Machine FREE BONUS 10 PIECE ACCESSORY KIT: * Straight Nozzle * Bent Nozzle * Scrubbing Brushes * Round Nylon Brush * Squeegee * Straight Brush * Measuring Cup * Funnel * Cloth Cover * Steam Blast Solution The Monster 1200 is a breakthrough in household cleaning that is the world's first deep cleaner AND sanitizer. It can deep clean ANY surface, eliminate germs and bacteria, and even deodorize and remove odors from your home, without using dangerous chemicals or time consuming handheld cleaners. Today you can get the Monster1200, 10 Piece Accessory Kit and Clean Blast Solution all for only 2 payments of $33.33 + $9.99 s&p. Or if you'd prefer, you can make one easy payment of only $66.66 + $9.99 s&p....

193. Lego toy methodology to build toolbox system

This idea benchmarks to the famous "Lego toys"...My idea is to design small compartments made of metal and to hold tools, some of these compartments have drawers.....The overall idea is that each of these small compartments can be joined side to side and up and down....The consumer can then begin to build themselves an awesome toolbox a few compartments at a time and size that fits there use...They can also design the geometry as they see fit, long, high, in a pyramid, square etc...very exciting geometries come to mind. The design of these compartments to be ergonomic...Form follows function said Frank Lloyd Wright....

194. Template to secure doors in typical homes

This idea designs an ergonomic template to secure the typical outside door locks in typical homes...The template is bought rough at the key portion area....The consumer takes it get there key cut geometry.....This template is placed at the top lock in a typical outside door as shown in the sketch.....The bottom portion of this template can vary around the knob lock...Shown is a simple tab on each side of the knob lock tang which flips vertically and horizontally...It shows a spring tab that secures the two walls around the knob lock....Now nobody can open this door from outside since the knob nor the top lock can work together......The nice feature is that anybody inside the home can set and unset this template, meaning kids too. Buy as many as you like for your home, they are made of biodegradable material.....

195. Tape measure with holes for temporary fastening

This idea places holes in the tape measure reel, at the end near the tip...The reason is to give the consumers places where they can temporarily nail down the tape measure in order to measure items more securely, longer and for reference measures...Please review the sketch...Exact location of the proposed holes can vary, but I propose three...These holes have tiny grommets to make them a little more rugged as nails will be going in and out of them......A secondary idea is to have the 90 tip flatten out for temporary nailing.....Now the consumer can also mark circles with this tape measure.....

196. Fashionable wrist worn tape measure

This idea designs an ergonomic tape measure that is worn around the wrist...This design is fashionable to boot....The user can forget about where there tape measure went, it's right there on there wrist....The tape reel could be narrow...The sketch shows a left hand wearing the proposed design...This design could also include a miniature L.E.D flashlight?....I really like this design and would wear it....Some tools grow on you and I feel this would be one of those.....

197. The two man lumber dolly

This idea is a two man shoulder dolly designed to carry loads of lumber at the hard to reach places on the job site....I noticed the construction works having difficulty placing loads of lumber in hard to reach places...Why not have them team up with braces and move more pieces of lumber so they can work continuously at that area longer, without having to run back and forth.....This idea benchmarks to the Shoulder Dolly Lifting System for Moving Heavy Objects.....Reference data on shoulder dolly".....""""Enables you to move bulky, heavy items in and out of your home or office without the hassle of a wheeled dolly, and reduces the back strain of lifting heavy items such as furniture and appliances. Designed with comfort and safety in mind, the Shoulder Dolly allows you to move heavy objects through difficult to maneuver areas while helping you to prevent back injuries. The Shoulder Dolly Lifting System allows you to move large, heavy objects through other-wise difficult spaces such as stairs, uneven ground, doorways, turns, and long paths. Because the lifted item remains upright, you can move around tight corners that would normally be almost impossible to maneuver. Clinically shown to drastically reduce the strain on back, arm, and hand muscles, the Shoulder Dolly utilizes the larger, stronger muscles of your shoulders and legs to do the lifting. Hand trucks, wheeled dollies, and carts can leave marks or tire tracks on your floor, and often you are left having to lift the object from them in order to go upstairs or go around corners. Bending, lifting, and twisting your body in this fashion can lead to serious back injury. With the Shoulder Dolly Lifting System, there is no need to pause or change positions - moving the heavy object is left to one continuous, comfortable motion. No more uncomfortable positions! No more dented doorways and walls! The heavy duty version is ideal for moving companies and delivery drivers, with a more vest-like design, thicker padding, and a wider and longer lifting strap. Constructed of a heavy duty cordura material, the Shoulder Dolly Lifting System features military style hardware that can be adjusted to account for your height, the size of the object, and how high off the ground you would like the object to be. These adjustments make it much easier for movers of two different heights to properly balance a large item. Moving can be stressful and tough on your body; with the Shoulder Dolly lifting System you can move bulky objects like dressers, entertainment center, hutches, sofas, mattresses, televisions, washer/dryers, and refrigerators in less time with much less effort and strain. ActiveForever is certain you will find that the Shoulder Dolly will help change the way you move"...

198. Tumbler lock with pictures instead of numbers

This idea is to design ergonomical tumbler locks for the construction industry using pictures instead of number for the tumblers...It looks like Wordlock has a patent with there letter tumblers so I thought to go one further...Actually humans think in pictures and so would be easier to remember than words....The exact pictures on the tumblers to be determined via psychology analysis.....Respectfully.....Reference data:..."Wordlock® Never Forget Your Combination The Wordlock® family of combination locks are the first combination locks that allows consumers to set their combination using words NOT numbers. These products are the next generation in personal security. Words are easy to remember, numbers are easy to forget and Wordlock® products are unforgettable. spacer All Wordlock® products are designed with the consumer in mind. Each product allows you to reset the combination to a word or a phrase that has meaning to you and is easier to remember. Resetting the combination is so simple you can change it anytime you want and as often as you like. spacer Easy to Set • Never Forget™ Wordlock® was chosen by the US Patent office and QVC as one of the top new inventions of the year. These award winning products have been shown on Regis and Kelly, Home and Garden TV, ABC news....

199. Flood alarm for dwellings calls phones when alert

This idea designs a home/dwelling water flood alarm system. This alarm system calls the phones programmed into it with a general alert preprogrammed voice message.....The sensor pads for this system can be portable in that the consumer could switch locations inside the swelling they want to watch out for such as kitchens and bathrooms. This is a wireless system...It's about time a flood alarm is offered to the general public at a reasonable price.....

200. Ergonomic neck chain for tape measures

This idea designs an ergonomic neck chain used to carry small tape measures.....The key in this design is ergonomic... This neck chain is designed as a separate item which the consumer buys and attaches to smaller tape measures, mostly for field use as not to lose the tape ...Some of us just don't like the belt clips.....

201. Straightedge with end clamps for straight cuts

This idea ergonomically designs a straightedge with integrated end clamps. This straightedge also folds up for storing in the back of a truck or trunk....The purpose of this tool as shown in the reference picture, is for the consumer to clamp onto a piece of work in order to give tools such as circular saws an alignment line........This tool is for the field and unusual cutting situations....

202. Software program simulates carpentry work

This idea is to design a CAD (computer aided drafting) software program to simulate all the workings of carpentry...This CAD is intended for the craftsmen and contractors ....In this proposed CAD program, the user would be able to do the following:. Pick out there lumber from the big box stores, cut it from a list of tools in a library, assemble it, pick hardware from a library of vendors, paint it, make changes...this program can simulate how a certain size nail reacts with a certain type wood, if the nail is located to close to the edge will the wood split....This idea can benchmark to CAD software program "PRO-ENGINEER 4.0" in which we design in sheet metal as it actually behaves with press brakes and the like......

203. Staple joins construction joists, studs, rafters

This idea designs a tool which staples two or more pieces of construction lumber around there perimeter...Used for but not limited to studs, floor joists and rafters....This tool secures the new staple to the lumber by press fit...This idea is for strengthening the joining of construction lumber in addition to conventional nailing....The staple can be three inches wide...The tool to be portable and easy to use...Now the contractor of high end homes has an extra method to secure framed lumber.....

204. Electrified wood framing staple keeps lumber dry

This idea design a new warming metal framing staple... It fits around typical building lumber, its 3 inches wide, it has an electrical system which warms it via metallic reaction......In turn this staple warms the air around the building lumber (floor joists, studs, rafter) and keeps them dry...The idea is simple, wood tends to react to it's temperature environment.....This new "staple" as I call it has an electrical wire running to a main heating unit inside the house..Perfect for high end homes....I hope this makes sense, otherwise I can make a sketch...Respectfully.... Reference data:....."Because of its hygro..scopic nature wood is always seeking to equilibrate its vapor pressure with that of the environment. That is wood always try to maintain its Equilibrium Moisture Content (EMC) which correspond to the environment. It is the unique characteristic of wood that expends when it absorb moisture if its EMC is higher than its existing moisture content, whereas contracts or shrinks when it loses moisture if its EMC is lower than its existing moisture content.....

205. Portable on site optimizing cutting systems

This idea benchmarks to the three new models of optimizing equipment for lumber cuts, as referenced below.....My idea is to fuse this optimizing equipment with the fact that contractors need a field portable machine heavy duty waterproof and indestructible machine which is portable, to actually save money at the source, which is at the construction site.....Respectfully..... Reference data.....""""OMGA T 421 OPT The T 421 OPT is a programmable cutoff saw for optimized cutting of wood stock and defect removal. Through a laser system, the operator can grade the board and mark the defects to be removed. The machine is equipped with a controller capable of performing five work cycle programs. The onboard computer is able to store cutting lists with lengths divided into two different grades and the relevant quantity of parts. Cutting lists can be input direct using the NC keypad or with direct serial connection to a remote PC. The system is capable of cutting material up to 5.5 in. thick and 12.5 in. wide and has a maximum feed speed of 200 fpm. It is available in 13- or 20-ft. useable lengths. Manufacturer/ Model Name/ Feed Method Maximum Maximum Scanning Method Brand Name Number Feed Rate Material Width OMGA T 1020 NC Auto infeed, AC servo 800 fpm 12 5/8" Fluor. crayon OMGA T 520 NC Auto infeed, AC servo 400 fpm 14" Fluor. crayon Mereen--Johnson Rip Navigator Mereen--Johnson "Rip Navigator" rip optimizing system electronically measures lumber gross width, optimizes the rip solution, positions the "Select" saw and activates the appropriate laser lights to guide the operator to the correct feed position while recording what is being ripped. This rip tracking system includes a Panel PC with 15.1--in. color touch screen, Ethernet port and modem providing integration to the office. Manufacturer/ Model Name/ Feed Method Maximum Maximum Scanning Method Brand Name Number Feed Rate Material Width Mereen--Johnson Rip Navigator Manual, semi 1 to 45 bpm 31 Multiple and fully auto technologies Dimter introduces the new OptiCut 200 Exact Available from Weinig, Dimter's new OptiCut 200 Exact crosscut saw combines the volume of throughfeed cutting and the accuracy required for finished dimension parts (+/- .5mm). The ability to process finish-dimension parts without additional machining, like double-end tenoning, greatly eliminates secondary handling, labor and costs. Manufacturer/ Model Name/ Feed Method Maximum Maximum Scanning Method Brand Name Number Feed Rate Material Width Weinig--Dimter Quantum 450 AC servo/ 1,470 fpm 11.25" Crayon marking/roller belt driven scanning Weinig--Dimter OptiCut 350 AC servo, 760 fpm 12"/ 17" 350 XL Crayon marking/belt driven scanning Weinig--Dimter OptiCut 204 AC servo/roller 590 fpm 8" Crayon marking/belt driven scanning....

206. Wood nail with disappearing head

This idea designs a wood nail with a special head that disappears after the nail has been driven...This is a play with material hardness...My idea is to leave the shank portion of the nail right under the head a little weak, but not so weak as to break off once the nail is being driven vertically down.....Once the nail is almost flush to the wood, the craftsman uses the claw side of the hammer to simply snap the head off, then drive the small portion of the nail down...I can see this being tangible and useful....

207. New biodegradable moisture barrier - construction

This idea designs a new moisture barrier for the construction trades....Traditionally Polyethylene is used but it is not biodegradable....This barrier will degrade after a few hundre years and not add to the plastic waste in our planet. ......Background on conventional Polyethylene...."""Polyethylene or polythene film is usually stable and resistant to degradation. Methods are available to make it more degradable under certain conditions of sunlight, moisture, oxygen, and composting. If traditional polyethylene film is littered it can be unsightly, and a hazard to wildlife. Some believe that making plastic shopping bags biodegradable is one way to try to allow the open litter to degrade. Plastic recycling improves usage of resources. Biodegradable films need to be kept away from the usual recycling stream to prevent contaminating the polymers to be recycled. If disposed of in a sanitary landfill, most traditional plastics do not readily decompose. The sterile conditions of a sealed landfill also deter degradation of "biodegradable" polymers. Polyethylene is a polymer consisting of long chains of the monomer ethylene (IUPAC name ethene). The recommended scientific name polyethene is systematically derived from the scientific name of the monomer.[1][2] In certain circumstances it is useful to use a structure–based nomenclature. In such cases IUPAC recommends poly(methylene).[2] The difference is due to the opening up of the monomer's double bond upon polymerisation. In the polymer industry the name is sometimes shortened to PE in a manner similar to that by which other polymers like polypropylene and polystyrene are shortened to PP and PS respectively. In the United Kingdom the polymer is commonly called polythene, although this is not recognised scientifically. The ethene molecule (known almost universally by its common name ethylene) C_2H_4 is $CH_2=CH_2$, Two CH_2 groups connected by a double bond, thus: Polyethylene is created through polymerization of ethene. It can be produced through radical polymerization, anionic addition polymerization, ion coordination polymerization or cationic addition polymerization. This is because ethene does not have any substituent groups that influence the stability of the propagation head of the polymer. Each of these methods results in a different type of polyethylene.....

208. Hammer with integral adjusting nail set tool

This idea integrates a nail set tool design inside the typical hammer head...The mechanism to turn the nail set tool in and out of the hammer is like a pipe wrench except with a tight tolerance design for smooth movement ...Now the consumer can nail down there nails and quickly set them under the surface of the wood with a turn of the wheel on the hammer head........reference data:...""""The pipe wrench, or Stillson wrench is an adjustable wrench used for turning soft iron pipes and fittings with a rounded surface. The design of the adjustable jaw allows it to rock in the frame, such that any forward pressure on the handle tends to pull the jaws tighter together. Teeth angled in the direction of turn dig into the soft pipe. They are not intended for use on hard hex nuts because they would ruin the head, however, when a hex nut becomes rounded beyond use with standard wrenches, the pipe wrench is sometimes used to break the bolt or nut free.

[1] Pipe wrenches are usually sold in the following sizes: 10, 14, 18, 24, 36, and 48 inches. They are usually made of either steel or aluminum. Teeth, and jaw kits (which also contain adjustment rings and springs) can be bought to repair broken wrenches, as this is cheaper than buying a new wrench…..

209. 90 degree handle on Max grip pliers

This idea takes the grip pliers and turns the handle 90 degrees in order to give a new ergonomic hold...This is a new product idea...I note in some situations such as working under the hood of a car the handle on conventional pliers is too long......

210. Metal strapping tool for rough frame walls

This idea is a strapping tool that uses reels of metal to re-inforce e the typical wood frame wall in rough frame mode and while still lying horizontally before raised into position...This tool straps the top and bottom sill of the typical wood frame wall, it secures it with this metal strap...The results is a super strong structure...This tool will also work on rough framed floors and ceiling...This tools reinforces the wood frame structure against earthquake and the like....Exact spacing to be determined by test-ing....Respectfully.....This idea benchmarks to shipping tension-ers used in shipping departments...Reference data:...Tensioners draw strapping tight around your packages. Tensioners for Open Seals— Use on flat surfaces. Strapping is wrapped around the package and overlapped on top. Tensioners for Closed Seals— Use for irregular surfaces as well as rounded objects. Place the seal on one end and fold strapping back to form a hook, slide other end of strapping through the seal. Tensioners push against the seal to tighten. (A&C) Feedwheel style have ratchet action for unlimited take-up. Tensioner holds the bottom strap while the feed wheel bites into the strap and tightens it. All have an adjustment plug for varying strapping thickness. (B) Windlass style hold the bottom strap and tension the top strap by winding around the take-up wheel. Ideal for creating extreme tension on the strapping. (D&E) Rack and pinion style have a limited take-up range. Strapping must be pulled tight manually before tension can be applied. Note: To create a compatible strapping system, select a tensioner that matches the width and thickness of your strapping or the "Strapping Group No." listed in the tables.........

211. Add unique clicking sound on tape measure reels

This idea is to add a unique clicking sound when the tape measure reel is pulled in and out of the enclosure...This idea benchmarks to the Harley Davidson motorcycle and the famous skipping cylinder sound....I believe a unique sound will add a bit of mechanical artistic flair the American consumer would like.....Human think in pictures and so a unique sound will be something people will remember from other tape measures... The exact sound and mechanism to be determined if this idea is appealing....

212. Tool to measure lead paint hazard

This idea designs an electronic machine that can read the parts per million of lead in typical home paint...These units can be sold to big box stores...The consumer brings in a sample of there paint, pay a fee at the big box store and have it tested, the take action as needed.......Reference:.."""Paint containing more than 0.06% (600 ppm)[vague] lead was banned for residential use in the United States in 1978 by the U.S. Consumer Product Safety Commission (16 Code of Federal Regulations CFR 1303). The U.S. Government defines "lead-based paint" as any "paint, surface coating that contains lead equal to or exceeding one milligram per square centimeter($1.0 mg/cm2$) or 0.5% by weight." [2] Some states have adopted this or similar definitions of "lead-based paint." These definitions are used to enforce regulations that apply to certain activities conducted in housing constructed prior to 1978, such as abatement, or the permanent elimination of a "lead-based paint hazard."""""...

213. Painters tray with locking wheels

This idea attaches wheels to a heavy duty paint tray..Two of those wheels are locking so it does not move anywhere while the user paints...When unlocked it can be pushed around the floor by the foot...How many times have you seen consumers push paint trays around the floor and cause a spill?...no more......

214. The Liquid nail gun

This idea ergonomically designs a gun which resembles a drill except this gun delivers metal paste into pre drilled holes in wood...This idea is so the consumer can drill a certain length holes for whatever reason needed.....The pre drilled hole has pours down the shaft , these act like little spikes for the paste metal to penetrate, and so gives is great gripping power...The gun itself can have a refillable tank and heater to make the paste... When the paste contacts air as it leaves the gun it dries in a few seconds...

215. Smoke & fire detector calls phones when alarmed

This idea is to hype up the typical smoke detector and give it the ability to call your the telephone numbers programmed into it when smoke and fire is detected by it...Perhaps it could also call the fire department with a pre recording stating "a possible fire at address..."....I note a lot of times these detectors go off when nobody is home...Why not have the detector call you directly?... The flip side to this idea is to have the detector set off the alarm in your car, that would get peoples attention.....

216. Microwave oven to dry lumber (studs,rafters etc)

This idea ergonomically designs a microwave oven which can hold wood pieces such as studs, rafter, floor joists etc...This oven has the same concept as your home microwave but the geometry is different in order to fit lumber sizes...The exact temperatures to be determined...Now the contractor has a tool to use in order to dry and cure wood quickly instead of air dry...If this idea takes off it can be used at the lumber clearing houses to also dry amounts of lumber quickly....This tool can be sold in different sizes to the size of the company it serves...

217. New reinforcing method to wood frame buildings

This overall idea is to reinforce wood frame buildings using steel dowels embedded into wood studs and rafters....In the end this system delivers a metal cage inside the wood frame to act against earthquakes and tornado's and the like...New tools will be needed to get this job done...The steel dowels are run in the middle and all the way down the typical stud or rafter...For example, a stud measuring 2" x 4" x 8' would have a dowel running 8 feet long...The dowel ends extend out of the studs or rafter son both ends to be welded to a metal connector on another stud or rafter...These dowels do not go in every single stud nor rafter but decided up by a civil engineering.....This idea benchmarks to masonry rebar systems for concrete and concrete blocks......

218. Work holding tool using robotic "Shadow hand"

This idea is similar to a previous idea except this time I use this "Shadow Hand" robotic arm to hold work in the typical workshop...This design is ergonomically designed like a mobile storage case... It has one or two robotic arms..The arm itself follows the movements of the human users hand who has an electronic glove that operates the actual robot hand.....For example the user would show the hand to grab a piece of furniture then lift it in the air and hold that position so the user can stain that piece of furniture....This is a robot and can lift relatively heavy loads so the uses are many..Another example is the auto shop, when one repairman wants to work on a tire in the air......Respectfully...... reference data:..."The Shadow Dexterous Hand is a humaniform robot hand system developed by The Shadow Robot Company in London. The hand is comparable to a human hand in size and shape, and reproduces all of its degrees of freedom. The Hand is commercially available and currently used by NASA, The University of Bielefeld and Carnegie Mellon University...."The Shadow Dexterous Robot Hand is the first commercially available robot hand from the company, and follows a series of prototype humanoid hand and arm systems.""....

219. Instant stopping of power tools with human body

This idea Benchmarks to "Sawstop" in which the human body acts like a circuit breaker to instantly stop a cutting wheel when any part of the body touches it..My idea is similar except adopted to power tool cutters and brings down the cost for the average consumer....This is the story of Sawstop:.....Gass started his experiments by running a small electrical charge through the blade of his Delta power saw. Whenever the blade was touched, the body would absorb some of the charge like a circuit breaker and immediately trigger the brake. He built a prototype, videotaped the demonstration and tried to license his invention to power-tool manufacturers like Delta. "One company said, 'We decided not to pursue this because the marketing guys say safety doesn't sell,'" he recalls. ...In the meantime, SawStop, which is available only as a commercial saw, will offer a less expensive version for hobbyists later this year. That may push more small shops to conclude, as Carl Seymour's did, that a safer saw isn't just good for workers; it's good for business. Gerald Wheeler, owner of Cabinet Door Shop, says two earlier power-saw accidents cost him $100,000 and two good employees, who suffered amputations. Seymour says he fixed himself up with "half a roll of toilet paper and a Band-Aid," leaving work that day with all his fingers, plus one really good story.....

220. Mobile storage units with self propelling feature

This idea is to add a self propelled feature onto the a mobile storage work stations....My idea is to benchmark to the Hoover commercial vacuum cleaner with this feature....I own one of these vacuum cleaners and can tell you first hand it's great, better yet on a heavy toolbox.....This is the spec on the vacuum:..."Self Propelled-A transmission driven by the motor helps to power the machine and loco mote it. Allows for forward and reverse movement with little effort on the part of the operator""".…"Hoover C1703-900 Commercial WindTunnel"...Imagine the users with there mobile storage units full of tools.....

221. Toolbox with Shadow Hand tool retrieval system

This idea designs the age of toolboxes...Picture in your mind a wall of different sized compartments, in each compartment there is some sort of tool stored...Now picture a robotic hand which can function exactly like the human hand...The user can have hundreds of tools stored and request via voice command to the robotic hand an automatic retrieval of that tool by the robotic hand.....Imagine your workshop with one wall dedicated just for storing tools..It would be very nice while working on a project to ask the hand to retrieve a certain tool stored on that wall...When the user is done this robotic hand also stores the tool back in it's exact storage location...Everything ergonomically designed of course.....Respectfully....Data on robotic hand..."""The Shadow Dexterous Hand is a humaniform robot hand system developed by The Shadow Robot Company in London. The hand is comparable to a human hand in size and shape, and reproduces all of its degrees of freedom. The Hand is commercially available and currently used by NASA, The University of Bielefeld and Carnegie Mellon University...."The Shadow Dexterous Robot Hand is the first commercially available robot hand from the company, and follows a series of prototype humanoid hand and arm systems.""....

222. Cable bike lock with individual locking tool hasps

This idea is a tool locking system when tools are stored in "soft storage tool bags"...This new cable lock has individual locking hasps tethered along the cable...The exact number of hasps to be determined per ergonomics....There is one universal key to open or lock all the hasps.....The idea is the end user can store there loose tools in the soft tool bag yet have the option of locking one or all of the tools onto this tethered locking cable system....In this day and age of so many pickpockets why not lock up each loose tool?....The idea goes the thief would have to steel the whole bag because this new cable lock is also secured to the soft bag with another locking hasp.....

223. The easy flip paintbrush

This idea designs a paintbrush with two heads and one or two knobs in the middle (opposite sides)...The paintbrush can be held in many ways but affords a more comfortable feeling per my testing with 20 men and women...This design allows for easy flipping of brushes for the work at hand...One brush could be thick brushes while the other side trimming brushes...The brushes can be replaceable to conserve on the actual handle knob design...I was thinking the handle could also telescope to allow the user comfort in distance from the paint...For example I cold be sitting down and paint with a telescoping handle...Things are comfortable with this paint brush...

224. Home fire sprinkler system

This idea ergonomically designs the typical building fire sprinkler system down to the home version....Maybe the sprinkler system can be camouflaged into decorative objects or household objects such as wall sockets...This system operates like any other in that smoke detectors and sprinkler heads sense temperature and smoke then turn on the sprinklers....This system could be first sold to high end builders such as the "Toll Brothers".....

225. Airplane black box for home & auto use

This idea benchmarks to the famous airplane black box except this version is a bit different meant for home and auto use... This ergonomically designed black box has a built in recorder which records continuously and disconnects upon some disaster/ accident such as fire or a crash...The object for this home & auto black box is to leave a audio recording of what the situation was that caused that incident...The user would know there is something recording them continuously and therefore could talk and leave evidence of the issue at hand...This black box is like the airplane version in that it is indestructible.....The flip side of this idea is to have recording devices in home and auto with the recording base unit at a separate site which stores the recordings in a super computer.....Respectfully....Reference data:.."""The flight data recorder (FDR) or Black Box is a flight recorder used to record specific aircraft performance parameters. A separate device is the cockpit voice recorder (CVR), although some versions (including the original) combine both in one unit. Popularly known as the black box used for aircraft mishap analysis, the FDR is also used to study air safety issues, material degradation, and jet engine performance. These ICAO regulated "black box" devices are often used as an aid in investigating aircraft mishaps, and its recovery is second only in importance to the recovery of survivors and victims' bodies. The device's shroud is usually painted bright orange and generally located in the tail section of the aircraft. It is also designed to withstand intense heat and pressure......

226. Hand held sweeping metals detector for lumber

This idea designs a metals detector for lumber....The idea is for the consumer to save there tools in not hitting hidden nails in wood...My idea benchmarks to the pictured "Wizard III" except is more ergonomic in design with wider detecting plane...I note no metals detectors like this in the big box stores?.....My benchmark reference:....The Lumber Wizard III is easy to use, locates virtually any type of metal and is very accurate. If you want to preserve your cutting tools, you need this metal detector!""......"All controls (top) are easily reached and self explanatory. You can use either the beep or vibrating warning - or both simultaneously. (bottom) The only adjusting necessary is fine-tuning the sensitivity."......"I was able to locate a single leg from a staple that was buried in this oak. I put it there for this evaluation but had difficulty seeing it the next day. The Lumber Wizard found it every time."......"The Lumber Wizard was able to see/find the fine point of this screw through just over 1" of solid oak! And, scanned from the opposite side, it did it every time"....

227. Folding home shop crane

This idea ergonomically designs a small folding crane which is very maneuverable and meant for the home shop user...The lifting capacity to be determined but less than 4,000 lbs...As we get older any tool which can help us lift or maintain weights afloat in the air is a big plus in my book....I note no small folding crane is offered at the big box stores....Let's try it...Respectfully.....My idea benchmarks to the "Troy ME3088 4,000-Lb. Air Hydraulic Folding Shop Crane"..Except my idea is a small version...specs: Lift the heaviest engine quickly and easily with just a squeeze of a trigger. Features: 6 control allows a single person to guide and lift the load. * 8 ton air hydraulic ram. * All steel construction. * Solid cast swivel wheels. * Large 62 extended boom. * Crane folds to a super compact size for easy storage and transportation, yet still carries the same load as the older style fixed cranes that took up so much room in a shop. Specifications: * Folded size: 33-1/2 L x 21 W x 58 H * Unfolded working dimensions: 64-1/2 L x 38 W x 99 H........

228. Recycled rubber for tool housings and handles

This is another Green eco-friendly product idea...My idea is to use recycled rubber for power tool housings and handles on the various tools...Rubber has many outstanding features that make it right use on tools...I'm talking about heavy duty tough rubber, as in tire quality...Heavy duty power tool housings in rubber would be flexible, tough, less likely to electrocute the user...The housings could also be exchanged for new ones after years of use..... On handles throughout various products the user would not even know they are holding on to a piece of heavy duty recycled rubber until they saw the sticker saying this is a "green product"..... Respectfully....Reference:.. ""Global Recycling Network is an electronic information exchange that specializes in the trade of recyclables reclaimed in Municipal Solid Waste (MSW) streams, as well as the marketing of eco-friendly products.""""".....

229. Oxygen bar system for workshop and home

This idea ergonomically designs an oxygen bar system for the workshop and home...Now the typical consumer can breath pure air for a few minutes per day..How about breathing pure oxygen after a long day in the woodust and paint fume workshop?.... Background info on oxygen bar:.."'The oxygen bar is a trend that started in the late 1990s in Japan, and quickly spread east to California and Las Vegas. Used for health and recreation as well'.........""""Oxygen Vending is the latest trend in oxygen bars. The units can be placed (stand alone) in health clubs (in front of exercise treadmills), health food stores, nightclubs, hotels, and retail businesses. The prices generally are $5 for for a 10 minute session and includes an aroma of choice. The reduced price (due to no personnel costs) allow clients to breathe recreational oxygen on a regular basis. The 10 minutes are an energy alternative to high priced cups of coffee.""""....."Oxygen Bar guests will normally pay $1.00 USD per minute to inhale an increased percentage of oxygen compared to the normal atmospheric content of 21% oxygen. This oxygen is produced from the ambient air by an industrial (non-medical) oxygen concentrator and inhaled through a nasal cannula or headset for a period of 5 to 10 minutes - or even longer. Medical grade oxygen through a nasal cannula typically delivers only a 33% concentration at a flow rate of 3lpm. While greater concentrations are possible via a nasal cannula at higher flow rates, these tend to be increasingly uncomfortable. Concentrations approaching 100% are only possible through a full face mask such as a non-rebreather mask at flow rates over 6lpm.""""....."Canned oxygen, a relatively new product, is a canned gas sold for inhalation. It typically contains only around 95% concentrated oxygen to avoid the problems of distributing medical grade oxygen. It may be flavored with flavors like "Mountain Mint" and grapefruit to make the experience more pleasurable. The addition of this product will supposedly give the customer a boost. The product is marketed as a healthy addition to the modern life, as a partner to purified water and natural food supplements. It is most widely available in Japan, where it is sold at 7-Elevens as well as given away as a prize in Japanese game shows."""".....

230. Add magnifying lens to flashlight

This idea is to add a magnifying lens a flashlights in order to be able to focus in and out with more precision....This idea enhances the typical flashlight....This idea benchmarks to the Dorcy flashlight:....."12 In. Aluminum measuring wheel comes with Ergonomic easy grip handle. The Snap-lock system provides durability in the field and collapses for easy storage. Extra Wide tire tread provides excellent surface grip. 85 Durometer wear resistant rubber. Heavy duty ABS Housing. Wheel is also equipped with a Kickstand. Ideal for Surveying Geological Sites, Land Measurements, Fencing, Paving, Road Construction, Accident Scenes etc. Measures up to 9,999 Ft.. 5-Digit counter with push-button reset.".....

231. Telescoping handle on hand held flashlight

This idea ergonomically adds a telescoping handle to the typical heavy duty flashlight...Sometimes we don't want to get that close to whatever it is we are lighting up and the beam is not powerful enough to light it up...Why not add as much of a telescoping handle so we can?......

232. Add bike kickstand to typical rolling measuring wheel

This idea ergonomically adds a bicycle type kickstand to the typical rolling measuring wheel....This is a natural addition....I do see people walk around with clipboards in one hand and this wheel in the other..When the user is writing there notes they tend to put this wheel down, why not let them vertically rest it with a stand?.....Respectfully......Reference:..12 In. Aluminum measuring wheel comes with Ergonomic easy grip handle. The Snap-lock system provides durability in the field and collapses for easy storage. Extra Wide tire tread provides excellent surface grip. 85 Durometer wear resistant rubber. Heavy duty ABS Housing. Wheel is also equipped with a Kickstand. Ideal for Surveying Geological Sites, Land Measurements, Fencing, Paving, Road Construction, Accident Scenes etc. Measures up to 9,999 Ft.. 5-Digit counter with pushbutton reset....

233. Wine makers hammer for general construction work

This idea benchmarks to the winemakers hammer...I note the golden ratio in the geometry of this hammer looks great, it feels perfect in the hand....Note the square head which gives more surface area to impact the typical small wood nail head...My idea is a similar hammer for the construction industry except to give a radius on the head in order to benchmark to a racing axe in which the swing of the hammer in combination with the round head gives a much more harder blow to the typical nail head....

234. Earthquake alarm via NANO sensors in footing

This idea is to design an actual sounding alarm that looks somewhat like a smoke detector which is placed inside dwellings....This sensor is alarmed via NANO sensors preset along the footing of structures...Since concrete is great for vibration resonance, then just a slight shake on the ground as the quake rolls in the soil will alarm the people inside the dwelling, giving them a couple of seconds to act before it actually begins to shake hard.....We just went through a quake here in California a few weeks ago...It would have been nice to have had just a little bit of warning via a special sounding alarm, to give me that split second to get under my desk....

235. Nail hammering aid to be sold with hammers

This idea ergonomically designs your best anti-vibe hammer handle with a round flat metal plate..The idea of this design is to help the consumer never miss hit a nail head because this aid is oversized and fits over the small typical nail head...The user would have this design in the left hand and the hammer in the right hand...The user would place the oversized plate of this new tool over the small nail head then strike this new plate which will drive the nail home.....Simple and smart tool....These can be sold with hammers...The consumer will very much appreciate this helping tool.....

236. Integral power down handle on utility knife

This idea designs an ergonomic handle bent forward atop the typical utility knife...With this handle the human hand can get a firm grip and be able to push down and backwards giving a harder force cut downwards....I notice the conventional way of cutting with utility knives is jut not enough with the human hand pushing down behind the blade....This design is an integral part of a new utility knife, but I was thinking it could also be an add on (sold separately) to conventional knives, if the handle is designed with an attachment feature....

237. Use ECOTRICITY co. to develop a home turbine

This idea is to use the green ECOTRICITY company to develop a home wind turbine machine for the United States......The purpose is for consumers to hook up there home work shops to these turbines and saving themselves money on the normal electrical bill......Respectfully.....Reference data about......Ecotricity - our logo Founders of the green electricity movement... That's what happens when you switch to Ecotricity. Because for every pound our customers spend on their Ecotricity bills, we spend a pound building new sources of green electricity....

238. Silencer for pneumatic nail guns

This idea ergonomically designs a gun type silencer to use on pneumatic nailing guns...The exact area of silencing to be determined......They just make a lot of noise, let's quiet them a bit.....

239. New nail gun tip prevents self shooting accidents

This idea is to ergonomically enhance/design the end of the typical nail guns (where the nails come out) in order to lessen the accident rate of consumers shooting themselves by accident.... My second idea is to enhance the nail belts for these guns, possibly using on dummy nail in-between the cartridge at certain intervals, much like they do in the military with tracer rounds in ammo belts....The question is: "DOES EVERYONE EVENTUALLY GET SHOT WITH A NAIL GUN'?...Here are some interesting comments:..."I have nailed my hand to a wall, had a nail bend when it came out and broke my thumb, shot myself in the leg with a 16d twist nail""....."I've been shot twice in the past 4 years. First time was toe nailing some handrail into a 4x4 post, not realizing there was a knot there, and it popped out the top and went an inch into my palm. The second time I wasn't doing the shooting. it was one of those contorted, strange angle situations, and my buddy shot me straight through the side of my left ring finger.

It didn't fracture bone"""""..."I also have the same finish nailer stories. Several 90 degrees and into a finger""""......"Finish nailer has gotten me 3 times over the years. In each case, the 2" nail did a 90 or a 180 and came out somewhere I wasn't expecting it. Just pulled them and kept working.""".....""I shot myself 4th day on the job as a framer. Nailing in fire blocks had the gun double fire and stuck it through my thumb. It did not hurt near as much as I expected. Went to the boss and said "Hey Randy... I just shot myself" He looked at it and so did the sparky on the job (sparky almost lost his lunch) Randy pulls out the pliers and pulls the nail right out. We bandaged it up and went back to work."""

240. Tin snips with offset handle to blade

This idea is to ergonomically design a tin snip which has the handle and blade offset from each other...The cutting action might be more comfortable than conventional flat plane snips currently offered....This idea benchmarks to a very ergonomically designed scissor from this company www.enjoyzibra.com/openit....please review website...

241. Tape measure with 90 degree measuring tip

This idea ergonomically designs the very end of the tape measure into a 90 degree turn in order to let the craftsman measure diagonal surface conditions with more accuracy...For example ,when boxes are assembled the craftsman will sometimes check for perpendicularity by measuring opposite corners on that box... This new 90 degree tip lets them set and measure more precisely ...

242. Hydrogen fuel cell powers tools of the future

This idea is to develop a hydrogen fuel cell applicable to certain if not all power tools of the future..."""Intelligent energy company"""".... is the world leader in this field and have developed a hydrogen powered motorcycle...Why not try this on power tools as an R & D project...I have no ties to this company...I feel a company that a big company should look into this...Please review this companies website..."WWW.INTELLIGENT-ENERGY.COM....

243. Hollow core wood nail delivers glue after entry

This idea ergonomically designs a typical wood nail except the shank is hollow and holds a small amount of wood glue...The glue seeps out the nail point once contact is made with the hammer and goes into the wood....The glue seeps deep into the wood for added holding power....Now the woodworking people can rest assure extra holding power is delivered with these nails.....

244. White noise diffuser for unacceptable work noise

This idea is to ergonomically design panels of various sizes with mini speakers that automatically offset unacceptable noise by emitting white noise.....Here in California we have a huge problem with noisy construction because we live in high density areas, especially around our beaches.....If the contractor had a tool which he cold put up at various stations on his job site then the outrageous noise would lessen....This idea of "white noise" and "sound masking" is feasible to apply for construction work.....My idea is to offer panel of various sizes for degree of jobs being performed.........Respectfully.....Reference material:...Sound masking is the addition of natural or artificial sound of a different frequency (more commonly though less-accurately known as "white noise" or "pink noise") into an environment to "mask" or cover-up unwanted sound by using auditory masking. This is in contrast to the technique of active noise control. Sound masking reduces or eliminates awareness of pre-existing sounds in a given area and can make a work environment more comfortable, while creating speech privacy so workers can be more productive. Sound masking can also be used in the out-of-doors to restore a more natural ambient environment. Sound masking can be explained by analogy with light. Imagine a dark room where someone is turning a flashlight on and off. The light is very obvious and distracting. Now imagine that the room lights are turned on. The flashlight is still being turned on and off, but is no longer noticeable because it has been "masked". Sound masking is a similar process of covering a distracting sound with a more soothing or less intrusive sound.......

245. Tape measure w/hand crank dynamo for auto reel

This idea is to ergonomically design a tape measure with automatic tape reel feed out at the press of a button...The difference in this design is that the power source is backed up with a hand crank dynamo besides the batteries, to save energy and help the GREEN movement........This is a copy of my previous idea relating to hand crank dynamos:........"This idea is to ergonomically install hand crank dynamo's where applicable to electrical tools...My ideas goes that if you can get a little bit of free energy to back up an electrical tools battery source then it's a worthwhile idea in relation to the GREEN movement...I sincerely believe the consumer would flip out and smile if they saw these Tools give all they have in creativity to save energy.....Here are some examples of the tools which could offer hand crank dynamos to save battery life.:..."True distance laser measurer".....The "Tripod flashlight" have ..."The intelli measure"...."Intellisensor stud sensor".....If a company really get wants to rock the world then you might want to try adding handcrank dynamos to help with actual power tools.....

246. Add 360 degree swivel feature on toolbox handle

This idea is to add a 360 degree swivel movement under the handles of your toolboxes...By adding the swivel the ergonomics of hand movement when carrying toolboxes is much more comfortable than the rigid fixed position handle....When the human walks they also swagger a bit, why not take this into account and add a comfortable handle that allows the toolbox to swagger a bit too....It has always been uncomfortable to me when walking with a toolbox....Let's add some ergonomics ....

247. Sawdust sprayer acts as bark for wood structures

This idea is to prevent decay and mold buildup on typical wood studs used in structures....My idea here is for a new ergonomically designed sprayer to spray wood dust with a light adhesive onto the wood structure after it's been framed...This wood dust would act as the bark on a tree and be the first line of defense against wood decay......Now the building industry has a new protocol and that is to spray on a sort of bark on there wood framed structures....I believe the high end builders such as the famous "toll brothers" would use this....Respectfully...... Background information.....""Trees have several mechanisms of resistance against decay fungi. Bark is the first line of defense. No stem-decay fungi infect through intact bark. ....Sapwood is capable of active response to invasion. Parenchyma cells in sap-wood sense the presence of the fungus and initiate a doomsday response. A terminal metabolism kills them, but results in conditions that are unfavorable for fungi. Chemicals limit the progress of the fungi. Second, in many conifers, resin is piped in to seal off the area. Third, the cambium responds to trauma by producing a very effective wall in the xylem at that point that often restricts an invader to the wood laid down before then. The wall may extend for some distance away from the invasion or wound...... Heartwood resistance is very different from sapwood resistance. Heartwood is dead and there is no active resistance. Instead, chemicals are deposited in heartwood as it forms by dying parenchyma. They render it more or less inhospitable to fungi. Species vary greatly in heartwood resistance. Redwood, cedars are very high; aspen, birch are very low. Nevertheless, every tree has at least a few fungi that have learned to live in its heartwood and cause heart rot. ....

248. Nail gun combines pneumatic & powder capability

This idea is to design an ergonomic nailing gun which offers both pneumatic and powder actuation nailing methods.....I note the market has either one or the other, why not combine into one tool?....The contractor will like the feature of light framing capability with pneumatic then be able to switch over to heavy duty nailing in concrete and the like.....Plenty of testing to be done but feasible concept.....Respectfully......Background on powder:...A Powder-actuated tool (often called "Hilti guns" or "Ramset guns" after two of the companies who manufacture the tools) is a nail gun used in the construction and manufacturing industries to join materials to hard substrates like steel and concrete. Also known as "direct fastening," this technology relies on a controlled explosion created by igniting a small chemical propellant charge, similar to the process that activates a firearm. Unlike firearms, powder-actuated tools come in either high-velocity or low-velocity types. In high-velocity tools, which are now illegal to manufacture and/or sell in the United States, the propellant acts directly on the fastener. This process is very similar to how a firearm works. Low-velocity tools introduce a piston into the chamber. The propellant acts on the piston, which then drives the fastener into the substrate. (The piston is analogous to the bolt of a captive bolt pistol.) A powder-actuated tool is considered to be low-velocity if the average test velocity of the fastener does not exceed 492 feet per second. Although high-velocity tools are now illegal to manufacture and sell, there are some (made decades ago) that are still in use in the shipbuilding and steel industries. Powder-actuated fasteners are usually nails made of high-quality, hardened steel, although there are many specialized fasteners designed for specific applications in the construction and manufacturing industries....

249. Ergonomic hearing protection for the worksite

This idea ergonomically designs hearing protection for the work site in construction...My benchmark study indicates the actual users are wearing ear muffs used at gun ranges...Why not design two separate pieces that are hung around the ear like the hands free cell phones designs?...Respectfully....Some comments on hearing protection:.."When it's a situation that requires ear protection, I use earmuffs. Go to your local gun shop for good earmuffs. They have better ones than contractors supply shops I've found"""...."I like the stethoscope type. Works much like earmuffs without the bulk""...."I prefer the muff type but will admit to not wearing them as much as I should. I do always wear them around large power tools like table saws, planers""....""Our biggest problem is around grinders and that type of stuff. Muff types are nice around larger "shop" equipment as you can hang them near to the machines"""...Background info:...Earmuffs are objects designed to cover a person's ears for protection. They consist of a thermoplastic or metal head-band, that fits over the top of the head, and a pad at each end, to cover the external ears. They come in two basic kinds: * Thermal earmuffs, worn in winter to keep a person's ears warm. * Acoustic earmuffs, also known as ear defenders: cups lined with sound-deadening material, like thermal earmuffs and headphones in appearance, which are worn as hearing protection. These may be carried on a head-band or clipped onto the sides of a hard hat, for use on construction sites. Some manufacturers combine headphones with ear defenders, allowing the wearer to listen to a music, communication or other audio source and also enjoy protection or isolation from ambient noise. Acoustic earmuffs were created in Italy in 1982....

250. Panel carrier for lifting oversized sheets (4 x 8)

This idea designs an ergonomic and cheap one handed tool which the consumer uses to carry sheets up to 4' x 8' ...This idea benchmarks to the GORILLA GRIPPER yet is cheaper and enhances the ergonomic design...It would be nice to offer an ergonomic and inexpensive tool the workmen uses to move large sheets...Respectfully.....Spec on the Gorilla Gripper:..Please also see there website.... "Gorilla Gripper" $49.95 Panel Carrier for gripping, lifting and carrying plywood and other sheet goods... from the top!....

251. Combine oil free and oil lubricated compressor

This idea is from my benchmark study to get the best of both worlds regarding compressors...I note the heavy duty work guys seem not to like the oil free compressors, it seems to have a negative stigma attached to the name...Although the science behind the oil free is great...Why not combine both technologies to advance this subject and offer an oil free tank with oil lube air?....I believe this to be a winning ingredient.....

252. First 12" of tape reel to act like rigid ruler

This idea is to make the first twelve inches of your tape measure reels into a sort of flexible flat ruler in order to let the consumers use the edges to mark off items...I was just noticing today the guys in our shop using tape measure edges to mark off items before cutting...Our guys struggle because the tape measure reel edge is not flat and not rigid at the first 12 inches or so...A slight design modification to your reels will sure come in handy to us...We have hundreds of guys working here...Perhaps thicken the first 12 inches of the reel and make it flat to act as a ruler........

253. Miniaturized metal detector onboard circular saw

This idea is to design an extended plate in front of the circular saw cutting wheel which also comes in contact with the wood being cut and integrate a miniaturized electronic metal detector on that plate. This metal detector senses objects such as nails to the depth of the cutting blade on the saw....The idea is for the detector to shut power off on the saw once a metallic object is censored in the pathway of the cutting wheel and so saving the cutting wheel from damage....

254. NANO GPS security system for tools

This idea is a tiny device ground positioning satellite alarm system which can attach or be integrated into high end tools... The purpose is when the user sets the alarm they are notified when the tool is being moved...Stolen tools is a big deal in our country, let's try to help the consumers, especially the contractors....This idea benchmarks to Dewalt's "MOBILOCK" except the Dewalt version is to big...Respectfully.....Reference material about the moblillock...:MOBILELOCK can be used just about anywhere and across several industries. Some of the most common applications include:....Construction Security The device can be hidden on construction equipment to protect some of your most valuable assets......Law Enforcement As either a motion detector or tracking system this device can be an instrumental tool for law enforcement.....Cargo and Shipment Tracking.MOBILELOCK can be used to monitor motion inside containers and to track your shipments throughout the country. ....Remote Security Rest assured knowing your vacation home or boat is safe using the MOBILELOCK alarm and location system.....

255. The one man joist hanger hand tool

This idea is a hand held tool which helps the typical contractor set joists on rafters and beams or posts before securing down using only one man when the load is not over bearing....This product idea benchmarks to the product JOIST JAW as pictured in this attachment...My idea is a more ergonomic design with handles and improved locking design...If you are able to watch the video associated with the existing JOIST JAW then you can appreciate this tools functionality...

256. The nail selector calculator

This idea is to design a calculator which the consumer uses to help them find the exact type of nail for the job at hand...This tool looks like a calculator and gives a digital readout of the answer...A great tool for the contractor and busy Do it Yourselfers.........Here are the questions this new calculator can answer for us:........To determine the type of nails you need for a project, ask yourself a few questions. What are you going to be joining together? Flooring, roofing, and framing jobs all call for different types of nails. How long a nail is required? In other words, what is the thickness of material to be joined? Are you going to use hard or soft wood? A good rule of thumb: for hard wood, ½ of the nail should end up in each piece. For soft wood, the receiving piece of wood should receive 2/3 of the nail. Lastly, will the finished project be indoors or out? If outside, exposed to the elements, consider using galvanized nails. Coated with zinc either by dipping or electroplating, these nails are rust resistant.. Coated nails have an adhesive coating applied. As they are driven in, the coating heats up from friction. As the nail cools, the coating hardens and bonds to the wood. A non coated nail is referred to as bright. Nail length is indicated by its penny size, abbreviated d; thus, a 2d nail is referred to as a 2 penny nail, a 4d is a 4 penny, and so on. As the penny size increases, the nail increases in length and thickness. The most widely used nail is the aptly named common nail. With a flat head and barbed shank, this nail has excellent holding power, and is used for construction framing and other general fastening. Larger framing nails may have a textured head to prevent hammer slippage causing injury or damage. A variation of the common nail is the duplex or scaffolding nail. It has a double head, one stacked on top of the other about ¼ inch or so apart, and as it's name suggests, is used to build scaffolding, scenery, and other temporary structures. The nail is driven to the first head. When the time comes to remove the nail, the second head is used to pull it out. Finishing nails and casing nails have similar shanks as the common nail, but the heads are quite different. The heads are smaller and able to be indented into the wood surface. When driving a finish nail, stop when the nail head is about 1/8 inch above the surface. Then place a nail set into the finish nail's small indentation in the head and drive the nail under the surface. The small hole in the wood's surface is then filled with putty. Casing nails are a bit heavier than finishing nails, and are used in cabinetry and trim work. Brads resemble finishing nails, though thinner, shorter, and smaller. They are used to attach molding to walls or furniture. A cousin of the brad is the panel nail. As their name suggests, they are used to affix paneling to walls. They differ from brads in that they have rings around the shank for extra holding power......

257. Radius back handle for hammers

This idea is a hammer handle with the radius going backward ..This idea benchmarks to certain outdoor tools which have a curvature backwards for a unique swing action...I am thinking of trying this curvature on a hammer too...This has to be tested for exact radius on handle.... The swing action is awesome on radius back tools.

258. Heavy duty contour work pillow

This idea is an ergonomically designed heavy duty pillow designed like a tear drop and sloping from high side to low side... The material is a super duty nylon with contour space age material inside to cushion the user...With this design the construction people have a rest position pillow to cushion them in there work...It can be used as a kneeling pad, a sitting pad, an arms cushion and a back brace, to name a few positions.....

259. Ergonomic and inexpensive drywall tape dispenser

This idea is to ergonomically design a cheap drywall tape dispenser resembling the typical tape dispenser sold in stores...I could not find a reasonably designed tape dispenser for drywall tape....I found these comments about this topic.... Respectfully..:.....Hey do any of you seasoned finishers have any contraptions to hold and dispense drywall tape. I'd love to see some ideas and pictures would be real nice.... Paul, I remember a tool called a 'shoe'. The roll of paper mounted to a spindle on the top and there was a hopper that held a small amount of mud. I worked much like a caulking gun, squeezing the handle would produce a thin ribbon of mud under the paper. It was a two man job, one ran the shoe and the second guy worked the tape and excess. I guess that it was not all that successful as I haven't seen one in over 30 yrs. You can't solve you're problems with the same level of thinking that created the problems. Albert Einstein ....Are you talking about the type of equipment called an auto-taper where the tape and mud are applied at the same time, the kind where you have to pump the mud into the thing before you start? Or just something that helps you roll out the tape and mud a bit easier like a banjo? Those auto-tapers run about a $1000 I think......I guess the Banjo is close but more than I want. I don't really want to dispense the mud just an easier way to handle the tape. I don't finish every day so the auto tapers are a bit overkill. I will think more about a Banjo. Thanks.......I still use a banjo to run tape, never could justify the cost of a bazooka, which run more in the $1200 range to start. The absolute newest and greatest thing in the world of taping tools is the Alpha-Tech System, this baby takes all of the hard work out of taping & finishing. This uses compressed air to push the mud thru your tools, about $5K for the complete system last time I took a look at them. They call it a CFS, continuous flow system http://www.apla-tech.com/cfsystem.htm I've also see the wall mounted dispenser that runs the tape thru a hopper to apply the mud. Never used this set up but I have seen pictures and it would be ok if you had multiple crew members running tape on a job.......

110

260. Air jet system to clean work on hand power tools

This idea is to ergonomically add a NANO air jet system on-board hand power tools in order for the air to clean the work being done, example around the chuck on a drill,,,,This idea benchmarks to stationary drill presses which have the nozzle near the drill chuck to air clean the work...This idea just brings the air cleaning down one notch to hand power tools....

261. Time reading tape measure reel.

This idea looks like the typical tape measure except the reel has time increments in minutes...The consumer can pull out the reel for two hours as example and have the tape measure pull in the reel one minute at a time...This gives a visual marker to the user on the field of how much time is left or allocated....Again the reel on this tape has time measurements...The color of the reel to be dynamic ...The user can hang this tool anywhere...The user decides on the time he pulls out of the reel...This design is a sort of visual marker clock...

262. NANO electrical surge protector on power cords

This idea integrates a NANO electrical surge protector right at the power cord on power tools.....Now the consumer can rest assured that the tool will never be damaged due to electrical surges.....Respectfully.......Reference on surge protectors...:A surge protector is an appliance designed to protect electrical de-vices from voltage spikes. A surge protector attempts to regulate the voltage supplied to an electric device by either blocking or by shorting to ground voltages above a safe threshold...

263. Pull out electrical cord on power tools

This idea ergonomically designs the power cord into the tool itself with a pull out feature like a tape measure......The new power cord is very narrow with ability to be wound up...Now the user has the feature of a neat storage for the cord as well as protecting it.....

264. Small diameter on power cords for flexibility

This idea is to significantly decrease the diameter of the typical electrical power cord on tools and give the user more flexibility when handling them.....This has been a lifelong problem, even tripping over the thick power cord...

265. Detachable power cord on power tools

This idea is to design an electrical power cord which is detachable from the power tool itself....This new cord has a locking feature which makes it impossible to accidentally release it without the owners knowledge....This idea will add a much nicer storage capacity for power tools...It also gives the consumer the ability to exchange worn out power cords instead of the whole tool....I have always hated dragging along the power cord into storage of the tool....

266. Self standing feature on tools

This idea goes through the tools and pinpoints tools which would be nice to have in a self standing up right position....For example it would be nice to have a hammer which stands up on it's own..Power tools too...Smaller hand tools like screwdrivers too...etc.....

267. Ecology friendly tools packaging - ECOLEAN co.

This idea is to minimize environmental damage by using ecology friendly packaging solutions...I have found this company called ECOLEAN which specializes in this ecology packaging solutions...I have no ties with this company but I was extremely impressed with there presentation on packaging...So my idea is to package all tools and anything else which is applicable with ecology friendly materials.....Plastics waste is messing up our planet...Company's like this should be highlighted whenever possible..

268. Interchangeable hammer head tips

This idea ergonomically designs the ability to change the very tip of the hammer head which strikes an object...This ability saves the consumer from having to buy complete hammers, saving them money....I believe the consumer will like the flexibility of changing hammer head tips.....

269. Add clocks on tools

This idea ergonomically adds clocks (digital or analog) at out of the way positions on any given tool in order to give the user the time of day..The technology exists to house these clocks in heavy duty cases, free from danger of being broken...Now the user can concentrate on there work and keep track of the time... Believe it or not people are always going for there cell phones for the time while on job sites...Let's help them a little bit with the time of day readily visible on the tool.....

270. Sure grip wood nail with tangs and pincers

This idea adds two tangs with pincers to the typical wood nail..The reason for the tangs and pincers is to help the nail stay down in situations when the hole of the nail shaft becomes enlarged due to factors such as earthquakes and normal expansion and contraction of woods...The tangs and pincers are thinner gage steel and are an integral part of the nail....this is not an add on...Throughout my life I have noticed nails becoming loose due to the hole enlargement factors....Now this design feature helps the wood nail grip longer....

271. Work boots with integrated sneakers

This idea ergonomically designs light duty sneakers inside new work boots.....These sneakers a part of the boot and slip in and out of the boots..The zipper design on the new work boot allows the user to slip inside the boot and actually fit there feet inside the sneakers... Saving time on the tedious task of taking off boots and refitting another pair of shoes....Doesn't seem like much unless one has been on a job site...Let's set a new standard and save time....

272. Tumbler watering tool for concrete work

This idea ergonomically designs a watering tool used primarily for concrete forming work. This tool is made of heavy duty plastic. It has a long handle which has a garden hose connection to receive flowing water...The other end of this tool has a series of hand positioned tumblers, each with a knob which opens and with closes the water stream....Now the user can open or close as many water jets as they wish at whatever positioned they wish to 360 degrees...I was observing concrete forming work the other day and noticed the men using simple garden hoses which they struggle with and look stupid, they were in the hosing down of the concrete form right after pouring....I noticed they have no real tool to water concrete form work for long continuous paths...With this tool they can design there water stream then walk along long paths of concrete form work with ease....This tool gives an even stream of water at any angle desired....

273. Work glove with rigid & non-rigid NANO feature

This idea uses NANO technology to design a micro cage inside the typical workman's hand glove...The NANO tech is battery powered and when turned on the glove structure becomes rigid at whatever position the user has set there hand / fingers to...Now the this rigid structure acts like a locking tool and also relieves the user from strain...Now user with some hand problem can use this tool and feel secure about there hand position... No slipping when these gloves are turned on because the glove is secured around the wrist...When turned off the user has a normal acting glove...The hardening skeleton inside the glove to be determined in NANO tech...Many uses for this glove besides the people with weak hands....

274. Kneeling pad with lip, handles & folding legs

This idea is a rectangle pad with thick rubber cushion....This pad has a lip around three walls..It has handles and has folding legs...This idea benchmarks to KOBALT and there kneeling pad with handles...My design enhances the KOBALT design with the folding legs and lips...Now the workman has a cushioned knee pad with a lip protection around three walls with the option of folding it open for sitting....

275. Transportable wood dust binding machine

This idea ergonomically designs an affordable machine that will take wood dust and compress it into formed shapes...This machine might be something that is towed to the job site but not too big....This product wood be for the contractor..Now they can take saw dust and reshape to use on the construction site..... With this machine we can recycle wood dust and so help the GREEN movement....There are different ways of compressing wood dust so some experiments must be done....Respectfully.... Here are some notes about compressing wood dust...... A wood dust briquetting machine from China that handles 80-120 KG an hour cost about $5000. It compresses and heats the wood to form round "logs". Very effective. Burns nicely. You can get an import broker and have them help you import it for about 8-10% of the total price. We got ours from Andy Wang in China. His email is tmcwoodworking@globalsources.com. You will find he is easy to communicate with and very responsive......... Check out pelheat.com They offer a complete set up for about $50,000. Still kind of pricey. They are in England. A lot of information on the site too......... Hi all, I am looking for a domestic solution. The commercial options are way too expensive. I cut and split my own wood for a log burning Rayburn cooker and for a log burning stove. I generate a moderate amount of sawdust (mostly oak, chestnut and silver birch) and would like to find a cheap/free method of turning this into blocks for burning on either stove.......

276. Lighting coil benchmarks to coiled air hose

This idea simply attaches one hook on each end of the typical coiled air hose designed except on this design it's not for air but for lighting...the coil is lite up via a modern method like using L.E.D.'s...The coils can be colored...Now the DIY and construction people have a great portable lighting system that stretches and shrinks for portability...Rooms upon rooms can now be lite up at will...Lengths of coils can vary...hooks at each end can vary too...Power source can vary from electrical to rechargeable battery pack.....

277. Use carbon fiber and aluminum on power tools

This idea is to experiment with carbon fiber and aluminum for structures on power tools....The results might be lighter and more durable power tools...One cannot pass up a great idea or modern materials... This idea benchmarks to the Porsche design studio and Metabo...Please read below and review picture..... respectfully.....Porsche Design have designed everything from tennis rackets to piano's, they have now teamed up with the German power tool manufacturer Metabo to create the Porsche Design P'7911 Multihammer Power Drill. The Porche Design Multihammer Power Drill's shell is made of carbon fiber whilst its structure is made of aluminum making it significantly lighter than other power drill's. The drill can easily drill through wood, tile and concrete and many harder materials with different drill bits......

278. Silicone trivet work gloves

This idea designs a set of work gloves with the silicone trivet material that is used in the kitchen...The reason for benchmarking to this material is for the heat resistance factor and the use in opening tight items like jars...Now the user working with power tools has cooler feeling hands and a sure grip while working with hot power tools for extended lengths of time......Reference below regarding the kitchen silicone trivet.....Bring versatility to your kitchen with the OXO GoodGrips Silicone Trivet. Designed for use as both a Trivet and a Pot Holder, the Trivet is made of 100% high grade silicone, which is heat safe up to 600°F and flame, steam and stain resistant. Whether you are protecting your hands or your dining room table, the OXO Silicone Trivet absorbs the heat. Our circular design features a unique rib pattern for improved flexibility and grip, a thick rim for easy pick up and a convenient hanging hole. In addition, the ribs are offset on each side to allow the product to spring open when used as a Pot Holder, yet lie flat when used as a Trivet. OXO's Silicone Trivet is 8-inches in diameter, matching up to the contour of your hand and large enough to sit underneath most pots, pans and casseroles. The Trivet is dishwasher safe and machine washable. Features and Benefits * 600°F heat resistant silicone protects hands and surfaces from heat * Silicone is flame, steam and stain resistant * Rib pattern enhances flexibility and non-slip grip * Use as a trivet, pot holder or jar opener * Thick rim for easy pick up * Machine washable and dishwasher safe * 8-inch diameter.......

279. Safety cone with integrated laser pointing up

This idea ergonomically designs a safety cone in red or yellow color which is used on all job sites...This new design installs a laser which is human friendly and has a colored beam.....The placement of this laser is in the middle and inside the cone and faces upward and through the hole opening at the top of the cone.....Now there is an added feature to the safety cone.....The people who do not see the safety cone below will surely see the laser beaming up....

280. Paintbrush w/ paint holding handle & delivery

This idea ergonomically designs a new paintbrush with a clear see through handle and which holds paint and is refillable ..This handle has a pumping action button for the human finger so the paint can be pressured into tiny holes and opening mechanism near the hairs and to wet the hairs on the brush...Now the consumer can have many of these paintbrushes filled with different colors to make the job that much more fun and easy.. The handle on this design unscrews whereby the consumer can attach a garden hose to flush out the paint and leave this tool clean...Let's try it......

281. Toolbox with fixed bottom and revolving top

This idea is an ergonomically designed heavy duty toolbox with a fixed bottom portion drawers section....The top portion of this toolbox can rotate 360 degrees so that each face (side) has a new set of drawers...Now the consumer can store more items while organizing them a lot better...The number of tiers in the bottom and top portion can be determined by marketing....

282. Solar lamp for construction job site

This idea is an ergonomically designed solar lamp specifically for the job site, it's portable..The overall height could be eight feet....The purpose for this lamp is of course to help save electricity when working in the evening and at night...This lamp is also adjustable in height so the handyman craftsman can use it at home as well...Rugged, beautiful and very functional what else can the consumer ask for?...I'm sure the consumer will love helping this world just one bit at a time....I do....Respectfully...... Background information on solar lamps.....With increasing interest in fuel conservation, the idea of employing sunlight for interior lighting has again become popular. The use of solar energy has received widespread attention in recent years as an approach to the conservation of energy. In recent years, solar powered lighting has become popular for lighting outdoor recreational areas and other remote locations around the home such as sidewalks, patios and pool areas. Solar lamps have a battery that is charged up during daylight hours by solar radiation and which provide light at night in gardens and open spaces when required. The solar panel converts solar energy during the day time into electrical power to charge up or to maintain a charge in a battery power pack which supplies the lamp during the night as required. Solar powered lamp has been commonly used in place of the electrically powered lamp to illuminate pathways, yards, parks and other areas, for the purpose of saving existing fuel resources and protecting the environment from pollution.

Solar powered illumination devices utilize photovoltaic devices to charge batteries which, in turn, activate a light source contained therein, in the absence of sunlight, for illumination and/or decorative purposes. Such solar powered lighting devices are desirable because they are relatively inexpensive and require very little maintenance.....

283. Electrical cord surface/floor standoff

This idea is an ergonomically designed standoff that is attached to the typical power tool in order to keep the electrical cord off the ground a few inches or more...These standoffs are attached and detached at will by the consumer and they are a few inches apart along the electrical cord....This idea can get more elaborate in that a chain of these standoffs can be attached to an electrical cord and when the user shakes the cord these standoffs pop open.....These standoff are round with spikes protruding, they press fit onto the electrical cord..No matter what angle the cord is at these still work as standoffs from a surface.....Let's face it none of us like to see our new power tools electrical cord mashing through the floor with dirt and water etc...Let's keep the cord clean........

284. fluorescent work shirt w/ walkie talkie pockets

This idea ergonomically designs a new work shirt for the job site....This shirt is padded to prevent bumps and bruises. This new shirt comes in long and short sleeve. This shirt has a special pocket near the heart to house a walkie talkie so the user can leave it in the pocket and still talk...This shirt is made out of the latest material for hot and cold comfort...The fluorescent color helps the user stand out on the job site....This idea benchmarks to hunting gear for outdoorsmen.......

285. Safety glasses with 24/7 L.E.D lights flashing

This idea is an ergonomically designed safety glasses for construction work..These glasses have L.E.D lights flashing 24 hours a day seven days a week (all the time)...The L.E.D's are placed on the front and sides of these glasses at the very corners near the eyes...Now we have an enhanced visual signal where the heads are of our coworkers on the job site. This idea is to enhance the safety standard...This idea benchmarks to a previous idea of mine called the "hard hat with L.E.D." ....

286. Tool for improving concrete curing method

This idea benchmarks to the product BAIR PAWS which is used in hospitals to keep patients warm or cool via a new hospital gown that has a blower connection to hot or cold air...My idea is to take this concept and to translate it over to help concrete cure perfectly once it has been poured. This new idea uses a blower and cover to maintain concrete at a constant temperature for at least one week in order to help it set perfectly......Respectfully.....Reference material below on concrete curing and the BAIR PAWS system:. Designed to replace traditional patient gowns and warmed cotton blankets, the Bair Paws® patient adjustable warming system is the world's first patient warming gown and adjustable warming unit in one system. Reference on concrete curing:... In all but the least critical applications, care needs to be taken to properly cure concrete, and achieve best strength and hardness. This happens after the concrete has been placed. Cement requires a moist, controlled environment to gain strength and harden fully. The cement paste hardens over time, initially setting and becoming rigid though very weak, and gaining in strength in the days and weeks following. Even though the cement reaction with water is completed over period of time (normally about around 240 minutes which is referred to as final setting time), the hardening of concrete and gain of strength is over a period of time 95% to 98% strength is achieved in 3 weeks or about 28 days. During this period concrete needs to be in a ideal conditions, controlled temperature and humid atmosphere, in practice this is achieved by spraying or ponding the concrete surface with water, thereby protecting concrete mass from ill effects of ambient conditions. The pictures to the right show two of many ways to achieve this, ponding – submerging setting concrete in water, and wrapping in plastic to contain the water in the mix. Properly curing concrete leads to increased strength and lower permeability, and avoids cracking where the surface dries out prematurely. Care must also be taken to avoid freezing, or overheating due to the exothermic setting of cement. Improper curing can cause scaling, reduced strength and abrasion resistance and cracking. ....

287. New Painters tray

This idea is an ergonomically designed painters tray with an integrated paint drip system and molded roller dip holder....This painters tray has paint already in the can..It has a handle..It has a molded holder for dipping paint rollers..It has a timed paint release system that can be opened and closed..When open the paint begins to drip along the roller dip tray...Now the consumer always has enough paint to wet the roller or brush..This tray can be carried via the handle and is refillable....

288. Wide mouth plane with handle and wheel

This idea is an ergonomically designed plane with handle and movement wheel at the rear...Now the consumer can have better controls over the hand plane...More finesse with this tool..Ever since I was a kid I hated the way the hand plane felt..Something uncomfortable about them...While shaving I realized how good my shaver felt in my hand and how well it moves around......This product idea benchmarks to the typical razor for men....

289. Circular saw cutter with reaming teeth

This idea benchmarks to a ream drill.....As the drill goes down it makes a narrow pilot hole in the surface being drilled then the drill widens out to the exact dimension wanted, this precision is called a ream hole...My idea is to design a circular saw cutter with it's teeth designed in the fashion of a ream drill, that is a pilot tooth cuts into a surface then as the cutter goes deeper it widens out....I believe this cutting action will be produce a more precise cut with less splinting on the material, especially wood.....

290. Tin snips w/ independent gearing force multiplier

This idea adds a small gearing mechanism in-between the handles and the cutting blades a the typical tin snips (shop scissors)...The reason for the gearing is to add a force multiplier to the cutting action via the gearing...Now we don't have to kill our fingers to snip tin....The has been a problem for as long as i could remember. The exact gearing ratio to be determined by testing.....

291. Toolbox made from heavy duty see thru acrylic

This idea makes a complete tool box, drawers and all, (except for handles and wheels) from see through acrylic...Now the consumer has an added visual aid when finding there tools.....I thought a change in the construction of a stool box would stand out in the "neat an cool" factor ...The construction method of this proposed toolbox would be like the stickley furniture company (tongue and groove, very neat joining methods)......Reference material...Respectfully....Acrylics are excellent as lens material. They are used in binoculars, cameras, and eye glasses. Acrylics are also used in stop lights and car headlights. Lighting fixtures, dishes, floor waxes, carpets, piano keys, beverage dispensers, and skylights all contain acrylic plastics......

292. Hybrid power tools

This idea is to challenge the lithium ion power tools from craftsman and go one step beyond...This idea adds a solar factor to power tools...To use NANO technology to recover solar energy to give a power tool a few minutes of use....My studies tell me that there is still too much environmental damage with batteries' helps...I believe the consumers will appreciate the green factor of solar energy into tools. I am not saying solar would take over electricity just a touch of it......Reference material:......The latest hybrid technology is the Plug-in Hybrid Electric Vehicle (PHEV). The PHEV consists of a gasoline-electric hybrid whose battery pack (usually Li-ion) is upgraded to a larger capacity, which can be recharged by either a battery charger hooked into the electrical grid or the gasoline engine (only if required). The car runs on battery power for the first 10 to 60 miles (16–100 km), with the gasoline engine available for faster acceleration, etc. After the battery is nearly discharged, the car reverts to the gasoline engine to recharge the battery and/or return the car to the charging station. This may get around the fundamental obstacle of battery range that has made nearly all pure electric cars impractical. Fuel costs (ignoring conversion costs), in principle, may be as low as 5 cents/mile. It's not clear yet whether converting an existing hybrid car will ever pay for itself in fuel savings. The biggest problem is finding a good, cheap, high-energy battery pack—the same problem that has plagued the pure electric car. If everyone plugged into the utility grid to charge up their car this would seem to be merely displacing the gasoline/diesel combustion problem to the typical coal powered electrical generating plant. But, if cars were recharged late at night this would allow the base load of the electrical system to be more efficient with a much more even base load and electrical power can also be generated by clean wind, hydro, tide power, etc. Since most travel is about 30 miles/day this may be the cleanest personal transportation system presently available. There is a "cottage" conversion industry for owners of existing hybrids, and several large auto industry groups (GM, Toyota, Mercedes etc.) as well as the US Department of Energy[19] are investigating this system. No major car company (as of late 2007) offers PHEVs yet. The typical "cottage" industry conversion car is the Toyota Prius (cost of conversion $5k-$40k), since it is a full hybrid with enough power in its electrical system to maintain typical city speeds......

293. Heavy duty butterfly net for construction work

Today I was observing construction people on a site, residential construction (apartment building construction)...I noticed these men picking up miscellaneous small items one at a time, very tedious work..One guy picked up ten little pieces of small 2 x 4 studs...This idea is to ergonomically design what looks like an over sized butterfly net, the length of a broom with a heavy duty net at the end...Now the construction people can pick up all sorts of items at a time..The user does not have to get face close to dirty items needing to be picked up...Has to be heavy duty netting at the end...

294. Fiber optics in concrete form work

This idea is a system using fiber optics to measure the settlement and movement of concrete form work..Now even the new home owner or consumer laying out concrete has a way to measure how well the concrete cured using this tool...Also great for after earthquakes have hit...The fiber optics would be embedded after the concrete has been poured but before curing so the fiber optics can lay inside....... Reference on fiber optics.....Even after a number of years of development, fiber-optic sensors have still not enjoyed great commercial success, since it is difficult to replace already well-established technologies, even if they exhibit certain limitations. For some application areas, however, fiber-optic sensors are increasingly recognized as a technology with very interesting possibilities. This holds particularly for harsh environments, such as sensing in high-voltage and high-power machinery, or in microwave ovens. Bragg grating sensors can also be used to monitor the conditions e.g. within the wings of airplanes, in wind turbines, bridges, large dams, oil wells and pipelines. Buildings with integrated fiber-optic sensors are sometimes called "smart structures"; they allow one to monitor the inside conditions and to gain important information on the strain to which different parts of the structure are subject, on aging phenomena, vibrations, etc. Smart structures are a main driver for the further development of fiber-optic sensors.....

295. Hand extension for construction work

This idea designs an ergonomic pick up tool to fit on the actual human hand with an extension of a few feet...In other words when you put this tool on your hand the dummy hand is three or four feet away but you can still manipulate it almost like your own hand...Now you do not need a ladder...Imagine all the tiny projects that can be completed very quickly with this tool...I'm thinking of an electromechanical mechanism that is battery operated...Now you can quickly touch up painting, nailing etc......

296. Laser beam safety zone on power tools

This idea shines a laser area around the area of the power tool which will be in motion and may cause harm to the user... The lights of the laser can vary but red and yellow seem to be the safety hazard colors...The proposed laser is human friendly just to give a visual marker for one to keep there limbs away from... This idea benchmarks to some sears craftsman power tools which have lasers but not safety lasers.......

297. Traffic crosswalk with new laser beam feature

This idea is to sell to cities...This idea installs a system of human friendly laser beams at each corner post of the typical street light intersection...These lasers are timed to go with the lights, they are about four feet from the floor...When a light gives the right of way to the typical pedestrian the new laser also turns on to hit the next pole and so gives the pedestrian another visual aid to signal people in there cars to hold back from crossing the laser....This idea is like street crossing lanes using lasers four feet off the ground...I just notice a lot of vehicles crossing the pedestrian lanes when waiting at red lights. why not give the pedestrians more help with a laser beam system.....I can make a sketch of this if not clear.....

298. Pretension reel on power tools for easy drop

This idea installs a pretension reel with handle on power tools..The reason for this idea is so the consumer can be working then grab onto the handle of this reel which gently drops the tool to the ground for a "soft" landing....Currently at job sites and Do it yourselfers the methodology is to grab the power tool by the cord and set the tool to the ground, the landing is usually "hard" eventually damaging the tool..When the consumer is working on stairs this is especially true...Let's be ergonomic and help the consumer give a longer life to there expensive tools. The reel is located conveniently on the typical tool.....

299. Dummy electrical cord on power tools to ease drops

This idea ergonomically designs a second electrical cord to cover the real power cord in the area where it enters the power tool..This cord is independent of the actual cord...This idea to to ease the drop when a consumer uses the cord to set the tool down on the floor...How many thousands of times have you seen consumers swing power tools by the electrical cord to set it down?.. especially when working at altitude...Why not design a dummy cord to sheath the real one..Now the consumer can swing and set down all the want without damaging the power tool....The actual length of this dummy cord could be 12 inches.....

300. Shock absorbers feature in power tool handles

This idea is to design a feature into the handles of the power tools to make them act like a shock absorber in a car...Power tools are dropped very frequently on the job site.....Why not integrate a shock absorber system into the handles to ease this drop...I'm sure the consumers will like the idea of saving there tools plus feeling a little better when they drop there tool....

301. Hammer saddle for handle to protect hand

This idea is a protector for the hand when it's wrapped around the hammer handle...This product press fits over the handle at two places and acts like a saddle....The material could be a thick piece of leather or new rugged composite....Now we can swing a little more freely without risk of breaking our fingers......

302. The two nail finger tab

IDEA TWO ONLY...This idea is a finger tab held on the index finger and holding two nails on opposite sides of that finger, mainly for wood nailing...Now the consumer can hammer down two nails at a time.....This tab disintegrates once the nails are hammered all the way down...This tab is like a positioning fixture for two nails...Different nail spacing's can be sold separately.....Now i can pick up two nails with one finer and still have my hammer in the other hand...Let's save the consumer time and hassle.....

303. Wood nail with slit tip

This idea puts a slit a the tip of a wood nail..When this nail is driven into wood each side of the slit tend to move outward and give more anchoring to the nail...These tips open out at an angle and so give a better grip in wood....

304. The three tip hammer

This hammer has three tips, it's designed to give the hammer blow to tip one a lot more force than a conventional hammer, because of the heavier head configuration ...Tip number one is the standard hammer tip...Tips number two and three can be mallet tips for lighter jobs....This design is inspired by the cave-man hammers where they would put a rock in triangle shape to a handle....The caveman might have had something there and so I 'm trying it again with a modern twist......

305. Earthquake damage results using NANO technology

This idea is to sell an electronic system using NANO technology so the consumers can detect how much damage was done to there structure after an earthquake........The small NANO chips would be placed facing each other all over the structure...After the earthquake hits, this system would measure the misalignment of these NANO chips and in doing so give a readout of those misalignments, interpreted via a software program which gives the owner a layman's overview of damage......These NANO chips are not limited to the structure, they can be placed all over the plumbing pipes.....Now the wondering is over, the consumer can have a readout of damage done to there dwelling after an earthquake...

306. Hand tools that can be used with one hand

This idea brings the ergonomic notion of being able to use certain hand tools like clamps with one hand instead of two..This proposed clamp could have a wrist attachment which secures it to the hand and allows the fingers to manipulate the tool, once work is gripped then make is easy to disengage the wrist attachment.....

307. Silicone pellet on wood nail tip

This idea attaches a small silicone pellet to the typical nail tip that will be used on conditions where wood splitting is possible...The pellet will lubricate the nail head as it's being driven into the wood and so helping to prevent splitting....Thank you.... Background material: Pellet is a generic term used for a small particle or grain, typically one created by compressing an original material. The generic term is used in a variety of specific contexts. When driving nails near edges or ends of hard, brittle, or knotty wood, you may have the misfortune of having a large crack appear, or even a piece of your lumber break out at the edge. There are a few ways of decreasing the likelihood of this happening. [edit] Steps 1. Blunt the point of the nail with your hammer. This is done by placing the nail on a hard surface with the point facing up, and tapping the sharp end with your hammer. The nail will then cut its way through the wood grain, rather than wedging a pathway. 2. Lubricate your nail. For especially hard woods like oak or maple, you may have better luck if you dip the nail in petroleum jelly, which reduces the friction of the driving process, and can decrease the probability of the wood splitting. 3. Avoid nailing very near the edge or end of the board. Where end nailing is necessary, angle the nail so that it can be started farther away from the end, but the nail will still get a bite into the other board it is being nailed to. 4. Use the smallest diameter nail which will do the job, or hold the work pieces securely. Obviously, a 16d or 20d nail, with its larger diameter, will exert more force on the wood's grain, making the wood more likely to split. 5. Predrill a hole slightly smaller than the nail's shand diameter. For a 12d nail, about 3/32 inch will reduce the pressure of the nail penetration without decreasing the nail's grip on the board. 6. Use wood with sufficient moisture if possible. Very dry wood is more likely to split, since the drier the wood is, the less flexible it will be. 7. Use softwoods where they are suitable for your purposes, rather than hardwoods. Douglas Fir, Southern Yellow Pine, or Lodgepole Pine are less likely to split than are Oak, Birch, or Maple. 8. Avoid nailing through, or very near knots in the lumber. Knots are usually formed by heartwood, which is harder and less flexible than sapwood. 9. Back any nail out if a crack begins to appear while driving it. A tiny crack is a sure indication the wood will split if you continue, possibly ruining a board if it is being used for trim work. You may have to look at choosing another location for your nail, further from the end or edge of the board, or predrill a hole for a wood screw or other fastening method....

308. Wood Nail with flex head for angle nailing

This idea puts some sort of weak point in the shank of typical wood nail, right under the nail head...Now the consumer is able to nail wood pieces together at angles which cause stronger bonds...Once the nail head reaches the surface of the wood it straightens out with hammer blows to be flush and therefore look neatly seated...The proposed weak point under the head can be an indentation or cut....Thank you... Reference: This is a tip from the famous Bob Villa ""Driving the nail at an angle may not be attractive, but will give you a stronger hold. Use a nail that is long enough to allow approximately two thirds of the nail to be driven into the thicker piece""....

309. Tool for counting exact pieces in stacked lumber

This idea is a hand held ergonomically designed tool the consumer uses to count the exact number of pieces in stacked lumber, mostly for lumber yards and big construction projects.....This idea benchmarks to the Braille Institute of America who have an experimental machine for the blind to read there mail...The blind take a picture of there mail and this machine readouts out the written information on the letter...My idea is similar except the software reads geometry, closed loop patterns which is what the side of a piece of wood looks like.....When the consumer goes to the side of stacked lumber they take a picture of it, the software would read the picture (all internally) of each closed loop pattern and give a count.... .No more standing there and counting huge stacks of lumber...Let's try it......

310. Tape measure with 45 degree measuring capacity

This idea is a tape measure with a unique reel in that it can turn at 45 degrees at certain intervals....This is a unique tool for measuring corners upon corners...The hinging feature for the tape reel allows for the 45 degree movements.....

311. Coat underside of tape measures reels w/rubber

This idea coats the underside of tape measures with a sprayed rubber lining for a higher coefficient of friction when rolling the underside of tapes against surfaces to give the craftsman a little bit secure feel....Throughout my life I have never liked how the tape measure slips to much on surfaces. I need a tape reel with a bit more holding power and no slipping......

312. Hammer with three position handle

This idea designs a new handle with three distinct positions for the user...Each position gives a different hammer blow...New and interesting.....

313. Hardware dispenser benchmarks to candy dispensers

This idea is an ergonomically designed small item hardware dispenser.... The main container which holds the hardware items is clear see through...A dial is designed so the consumer may re-trieve one piece of hardware at a time...This idea is like the candy dispensers we consumers are used to seeing...This dispenser is refillable and the consumer may put as many as they desire side by side to have a great looking inventory system..."KAN BAN" is the Japanese word for signaling system for low inventory levels.. We can add an advertising twist to this product with the word "KAN BAN" because the container is see through....I believe this is one of those products that is dumb in nature but consumers will actually use, its simple and needed badly.... Exact size limitation on hardware items to be determined.....

314. Electric stapler w/pinpoint shot location viewing

This idea adds ergonomics so the consumer may better view exactly where the staple on a typical staple gun will hit...I propose a protruding nose made from clear plastic and an angled front wall on the gun in order for the consumer to see where the staple will hit, viewing from top down...Notice how most people will look on the side of the staple gun to see the shot location... Even though staple guns do not have to be that accurate its nice to know one could give a better try.....

315. Tape measure with voice readout on measurements

This idea adds a voice readout for measurements on a new advanced tape measure...When a consumer slides out the tape and presses an ergonomically designed button it reads out that measurement, in whatever increment it's set too.....This is great for the consumer who doesn't have perfect eyesight, the contractor who needs a fast accurate measurement, when lighting is not so good in the area being measured and lastly confirmation of an accurate measurement......

316. Hard hat with attached laser tool

This idea ergonomically attaches a surveying or laser tool at the side of a new construction hard hat...A flip down window at the front rim of the hat allows the user to read data like a pilot would...Now the users hands are free for other tasks..A feature at the side of the hat allows the user to rotate his survey or laser tool....Tests to be performed to determine which laser tools are applicable for this....

317. Fire suit with special accordion hose & sprayer

This idea enhances the typical fireman's suit to include a water hose port on the back and two sprayer nozzles at the front..... In use the fireman arrives at the fire, it's determined the fireman must enter the building, a new accordion style water hose is attached to the fireman's fire suit via the new port at the rear... As the fireman enters the building the accordion hose is with him and is turned on so water flows through and the two sprayers at the front of the suit begin to spray water ...The fireman's hands remain free to handle other things such as dragging people out... The min idea is water being sprayed in front of the fireman suit as he determines the direction and situation...Since the hose is accordion like it will contract as the fireman back up...The hose connection is designed so the fireman can disconnect it quickly and run if necessary.....

318. Decibel readout power tools

This idea is simply to add a digital readout on power tools so the user knows how high the sound is. This would be a safety gadget but and useful over the long run for consumers..By adding this information the consumer will feel comforted in knowing somebody is doing what it can to maintain it's tools below the hearing damage level....This idea can also start a program for noise factors in power tools by better design and engineering...... Reference data: Environmental Noise Weakest sound heard 0dB Whisper Quiet Library 30dB Normal conversation (3-5') 60-70dB Telephone dial tone 80dB City Traffic (inside car) 85dB Train whistle at 500', Truck Traffic 90dB Subway train at 200' 95dB Level at which sustained exposure may result in hearing loss 90 - 95dB Power mower at 3' 107dB Snowmobile, Motorcycle 100dB Power saw at 3' 110dB Sandblasting, Loud Rock Concert 115dB Pain begins 125dB Pneumatic riveter at 4' 125dB Even short term exposure can cause permanent damage - Loudest recommended exposure WITH hearing protection 140dB Jet engine at 100', Gun Blast 140dB Death of hearing tissue 180dB Loudest sound possible 194dB OSHA Daily Permissible Noise Level Exposure Hours per day Sound level 8 90dB 6 92dB 4 95dB

3 97dB 2 100dB 1.5 102dB 1 105dB .5 110dB .25 or less 115dB Perceptions of Increases in Decibel Level Imperceptible Change 1dB Barely Perceptible Change 3dB Clearly Noticeable Change 5dB About Twice as Loud 10dB About Four Times as Loud 20dB Sound Levels of Music Normal piano practice 60 -70dB Fortissimo Singer, 3' 70dB Chamber music, small auditorium 75 - 85dB Piano Fortissimo 84 - 103dB Violin 82 - 92dB Cello 85 -111dB Oboe 95- 112dB Flute 92 -103dB Piccolo 90 -106dB Clarinet 85 - 114dB French horn 90 - 106dB Trombone 85 - 114dB Tympani & bass drum 106dB Walkman on 5/10 94dB Symphonic music peak 120 - 137dB Amplifier rock, 4-6' 120dB Rock music peak 150dB....

319. Two beam level/plumb with crosshair camera

This idea benchmarks to the ROBOTOOLZ 2 -BEAM LEVEL AND PLUMB CROSLINER Excepts it adds a crosshair telescope camera for pinpointing measurements.....Now the user can pinpoint lo-cations with a camera and take a picture of it if needed.....Thank you..... Reference material:roboToolz 2-Beam Level and Plumb Crossliner — Automatic Leveling, Model# RT-7715-2 Features super high-visibility horizontal and vertical beams, 140° verti-cal and 120° horizontal fan angles, and automatic leveling and out-of-level sensing. Independent or simultaneous (cross-line) beams for countless level and plumb applications. Unit can be taken out of self-leveling mode and locked in place to set any angle. 1-year limited warranty....

320. Long barrel extension for nail and staple guns

This idea is to ergonomically design a long shaft barrel tool which attaches to nail and another for staple guns....This long barrel is to help the consumer not have to get up close to the work needing nailing or stapling...Sometimes when a job is com-plete a few more nails or staples need to be added...Why not offer a tool which saves a few seconds of tedious nailing/stapling time...

321. Crowbar with scissor fulcrum lever for added power

This idea is a standard long nose crowbar with a scissor like bar with stand in the middle which folds open to act like a lever on a fulcrum....This scissor closes flush with the bar when not needed...Now the user has a longer lever arm in situations where more leverage is needed. The consumer opens the scissor arm in the middle of the bar, sits the base on the lever surface and begins to pry open whatever the job is...I found by my life's experiences with crowbars that sometimes a longer lever arm is needed for more power .... Why not design this longer lever right into the crowbar.....

322. Climate control inside sealed wood structures

This idea adds a separate climate control system for inside walls, attics and under floor...The purpose for this idea is to help preserve wood...This system maintains climates at stable temperatures so wood can last longer..Albeit a little bit more expensive for the consumer to have a separate climate just for wood preservation but they just might buy it on expensive homes.... My idea comes from the article below:....Respectfully Climate has an important bearing on the relative rate of wood decay, thus the expected service life of wood exposed to the weather. Researchers at the Forest Products Laboratory have devised the following climate index map to predict relative decay hazard regions in the United States. The map is based on the mean monthly temperature and number of rainy days. The most severe decay hazard location in the United States is the Southeast, where rainfall is high and weather is warm and humid. In the Northeast and Midwest, decay advances at a somewhat slower rate. In the Northwest, the decay hazard is moderate near the Coast but it can be severe on the Coast. Decay is less hazardous in most of the Southwest because this region is very dry. In mountainous regions, localized areas with marked differences in temperature and rainfall occur. Index differences due to this factor are not reflected in the map. Where climate is relatively uniform over wide areas, the map can be used with confidence. The climate index map primarily estimates the decay hazard of wood exposed above ground to weather. With certain restrictions, the map can also be used to determine the hazard for wood in contact with the ground. Any place where wood contacts the soil should be considered a high decay hazard, indicating that wood should pressure treated with a preservative. Homeowners, architects, builders, and marina operators can use this map for help in selecting the wood species or preservative treatment that will ensure maximum service life of wooden structures.....

323. Wood decay kit offered at home improvement centers

This idea benchmarks to the article below..A new test kit is available for the consumer to check if fungus is ruining there lumber on pre and post construction...The actual kit is pictured, a blue line indicates the results...This idea is to look into this product and offer it in home improvement centers..It shows how a company helps the consumer save money in preventing wood decay....respectfully.. Each year, billions of dollars are spent to replace wood products destroyed by decay fungi. A large percentage of this loss is incurred by consumers. Researchers at the Forest Products Laboratory (FPL) have developed a new testing procedure, the immunodiagnostic wood decay test, that detects decay fungi in wood prior to visible damage. Other than culturing and microscopic observation, reliable methods are currently not available for early detection of decay in structures. Early detection of the presence of wood decay fungi can prolong the service life of wood by preventing unnecessary replacement and will ensure that infected wood can be replaced or remedially treated to prevent recurrence of decay. Researchers drill a hole in a wood post This wood decay test is a rapid, inexpensive, and accurate way to determine if fungal decay is present in wood. The test detects the presence of wood decay prior to visible decay or measurable strength loss. It can be used by inspectors, millworks companies, structural engineers, and wood crafters in a variety of applications, such as * Building inspections * Historical restorations * Utility pole inspections * Structural maintenance programs (e.g., bridge decks, piers, railway ties, wooden aircraft frames) * Response to warranty inquiries * Efficacy tests of new preservatives * Quality assurance for lumber Comparison of negative and positive results using test cassette. This wood decay test is a one-step test that detects antibodies to decay fungi. Wood shavings from the sample are soaked in an extraction fluid for 2 hours. The extracted sample is added to the window of a test cassette. If the extracted sample contains fungal antigen, it binds to a latex conjugate and forms a complex that migrates forward on the test strip. Another antibody immobilized on the test strip captures the complex and forms a blue band that remains in the test cassette window, indicating a positive result. Prepared for: PATH - Partnership for Advancing Technology in Housing Prepared by: Forest Products Laboratory.....

324. New detergent to remove mold from lumber

This idea is a new detergent with electric scrubber to remove mold from lumber BEFORE it is sealed up on a typical construction job....The detergent is described below but not offered officially in home improvement centers...Maybe one can officially offer this detergent with electric scrubber...The exact formula to be determined for this new detergent....Thank you. REFERENCE MATERIAL:... Question:..What should I do about mold spots on lumber while building a home?... Answer:....The media attention to mold recently—and the extreme cases usually presented in those accounts—has made homebuyers especially concerned about mold. You can remove spots of mold with a detergent (or bleach) solution (and a little scrubbing action). The fact sheet, "Helping your Buyers Understand Mold during the Building Process,available at http://www.toolbase.org/ ToolbaseResources/level4BP.aspx?ContentDetailID=1182&Bucke tID=5&CategoryID=24, would be appropriate to review and possibly share with the homebuyers. After mold spots have been removed, allow lumber to dry (while stored properly), and be sure that the moisture content is less than 19 percent before enclosing lumber behind walls. You may need to talk with your lumber supplier about mold on lumber if it is a continuing problem....

325. Basic fun tool set for the handicapped

This idea benchmarks a hand tool set specially designed for the handicapped, with basic use of there arms and an IQ higher than 90...Designate this set "tools set for the handicapped"... Exact tools to be determined but should be the fun tools in woodworking....For example these tools can have a delay turn on mechanism of a few seconds to protect the user...Another example could be an anchored swing arm for the wheelchair to help the user move the tool around....Now the disabled can work in wood too.......Some statistics about the disabled respect-fully:.......In the United States, for example, Americans with disabilities constitute the third-largest minority (after persons of Hispanic origin and African Americans); all three of those minor-ity groups number in the 30-some millions in America. According to the U.S. Bureau of the Census, as of 2004, there were some 32 million adults (aged 18 or over) in the United States, plus another 5 million children and youth (under age 18). If one were to add impairments -- or limitations that fall short of being dis-abilities -- Census estimates put the figure at 51 million.....

326. Benchmark for new ladies hand tools

This idea is design a separate hand tool line specifically de-signed for women of all ages....These tools would weigh less and be reduced in size for easy management by the typical petite woman.....This idea benchmarks to existing company offering la-dies hand tools.....Please review this site: LADIESTOOLSONLINE. COM...This company offers pink tools sets ergonomically designed for women as presents.....

327. Folding stocks on heavy duty power tools

This idea adds a folding stock to your heavy duty power tools, especially the nailing guns...This is not an add on put a part of the tool design...This idea came from the military weapons, some modern rifles have folding stocks for ease of use...I feel that although the heavy duty power tools are light they should give the ergonomic opportunity to the user of being able to hold the tool with two hands, to split the weight after hours of use...I am a strong guy and would appreciate splitting the power tool load with two hands.....reference to attached picture: Side Folding stock with Pistol Grip, includes Free ribed Forend! Fits Remington 870....

328. Facial anti sweat gel for construction workers

This product is a body gel designed for the face only and to keep sweat away..You have to admit everybody has to stop and wipe there face when working specially the guys working at job sites.......Why not have a special gel for the face only to make the sweat drip around the face and at the same time have it feel like there is nothing on your face....It becomes a nuisance to continually wipe ones face..This idea benchmarks to the football players black stick to prevent glare and the following products mentioned below.......... What it is:An antiperspirant gel.What it is formulated to do:Bliss's mint-powered roll-on antiperspirant gel-now in a nifty new bottle-helps circumvent sneaky sweat stains and keeps underarms feeling fresh, cool, and dry.What else you need to know:On hiatus since 2005,................... United States Patent 4781917 Abstract: Disclosed are antiperspirant gel stick compositions substantially free of unbound water comprising from about 5 to about 50% of a solubilized antiperspirant active, from about 7% to about 35% of intermediate polarity emollients, from about 1% to about 5% of a benzylidene sorbitol, from about 15% to about 75% of a polar solvent, from about 1% to about 20% of a coupling agent and from about 0.5% to about 10% of a buffering agent. These antiperspirant gel sticks provide very stable antiperspirant compositions with good efficacy as well as excellent cosmetic anesthetics which are further characterized by their ease of manufacture. Also disclosed is a method for the manufacture of these gel sticks as well as a method for treating or preventing perspiration and malodor associated with human underarm perspiration.....

329. The drywall rake nailer

This idea is an ergonomically designed power nailing machine to nail drywall sheets to wood stud frames...This tool has protruding nailing mechanisms every 8 inches on center and 4 feet wide overall. This machine looks like a rake. Drywall is usually sold in size four feet by eight feet...Now the consumer can nail complete four foot sections of drywall to studs at once instead of one nail at a time.....Imagine this tedious task of nailing drywall to done much faster.....

330. Coca Cola can W/magnetic rim dispenser for hardware

This idea is a half sized coca cola can with a circular opening at the top that has a magnetic rim...Small hardware items like nuts and bolts and nails can be stored inside this can..When the consumer flips this can to one side a few of the hardware items inside magnetize themselves to the rim and thereby reveal themselves for easy retrieval....The cans are reusable and can vary in height.....This idea came out of me not liking the boxes that are used today to store hardware items, they breakdown or rundown, not cool in design either........

331. Nail holding tool at 45 degrees for skew nailing..

This idea designs a simple hand tool for holding nails at 45 degrees against wood stud walls that will be nailed into sole plates, this is called skew nailing....Everybody has issues with exact 45 degree nailing wouldn't it be nice to have a tool to help...This idea is similar to my previous idea for hammering nails except it adds a 45 degree wall...

332. Aluminum extruded building stud with wood inlay

This idea is a new interior wall stud used in the building industry...The new wall stud is extruded from aluminum with channels to slip fit smaller wood studs inside each end of the stud.. Now the consumer has the best of both worlds, a strong rigid frame plus the ease of working with wood for nailing, holes and boring...The exact design of the aluminum extruded stud is my secret.....The sizes of this stud are the same as conventional lumber,examples 2x4,,2x6,,2x8,,etc.....

333. Oversized dial and numbers on calipers

This idea is to ergonomically add an over sized dial with numbers on the typical dial caliper...If you look around all companies offer small dials and number on there calipers... Why not give us what we need to easily read measurements....The bigger dial will glide more easily on the finger tip. The over sized measurements will give the average Joe a better visual without squinting....

334. Broom with magnetic bristles - Magnetic sweeper

This idea looks like a broom except the bristles are magnetized to pick up metallic items on job sites...The bristles also act like a broom to sweep any other object as well...Now the DIY and contractor have a cool new tool.....Please see attached picture of what is currently on the market (ugly)....Reference data: AJC's New Mini Magnetic Sweeper is handy for a variety of jobs such as loading a nail bag, removing nails from gutters and cleaning metal debris from confined spaces. Features a spring loaded quick release and a 3.5" diameter stainless steel base. Overall height of unit is 6.5"....

335. Add the color blue to underwater tools

This idea is to add the color BLUE to the housings of your underwater tools....My reason for adding color BLUE is for better visibility underwater...The color yellow disappears at approximately 16 feet underwater...The color BLUE disappears at 98 feet.....Other notable colors are red which disappears at 2.5 feet...Green at 49 feet....By adding a more visible color then it goes a little bit more than the competition for tool alertness underwater....My data is scientific fact....

336. Space inspired tool enhancements

From my overview of Space shuttle tools I bring a few improvements that can be applied here on earth to our tools:...1... Enlarged openings and easy to push switches for working with gloves on hands....2..the handles of tools enlarged so they will take less energy to hold.....3..To keep control of tools on belts, each tool to have some sort of tether or locking system..4...A socket wrench has a key that has to be inserted into a holder before a socket can be installed at the end of the wrench. Once the key is removed, the socket is then locked onto the wrench and cannot be removed without use of the key again....

337. Benchmark to Gorillapod for tool holding/clamping

This idea benchmarks to the camera Gorillapod gripper except this version would be for tools...This is an add on tool for use with any tool...Now the consumer can stand or hang there tools temporarily anywhere...Please read reference material below and visit there website...respectfully.... The Original Gorillapod is the lightest and most versatile camera tripod available today. Featuring over two dozen flexible leg joints that bend and rotate, the Gorillapod will firmly secure your compact digital camera to virtually any surface — anywhere and everywhere! The Gorillapod is the ideal camera accessory for photographers on the go. Throw it in your pocket or backpack and you'll be ready for your next adventure! While the Gorillapod serves all the functions of a traditional camera tripod – steadying your camera under low-light conditions, taking timed group shots, etc. – it is the only tripod malleable enough to provide you with the perfect shot while wrapped around a tree branch, hanging from a pole, or perched on a jagged rock. The possibilities are endless....

338. Pneumatic tools to use Nitrogen instead of air

This idea simply notes that Nitrogen works better than air, why not change up future tools to use Nitrogen?..I feel the tools would work even better and last longer than now with less problems than now..Of course an infrastructure for Nitrogen has to be built but those are the growing pains of great tools....The benchamrk for this idea comes from the race care industry..Here is a short synopsis on Nitorgen.:...Many race car teams use nitrogen instead of air in their tires because nitrogen has a much more consistent rate of expansion and contraction compared to the usual air. Often, a half pound of pressure will radically affect traction and handling. With track and tire temperatures varying over the duration of a race, the consistency of nitrogen is needed. Nitrogen pressure is more consistent than normal air pressure, because air typically contains varying amounts of moisture due to changes in the relative humidity on race day. Water causes air to be inconsistent in its rate of expansion and contraction. So, a humid race in the southeast United States or a dry race in the desert western United States could make for unpredictable tire pressures if "dry" nitrogen were not used.....

339. Shot bag for construction work

This product idea looks like a small square pillow and made out of heavy duty cowhide. It has shot or sand inside...The purpose for this product is to suggest that not all carpentry, DIY or handymen around the house need hard surfaces to work on with there hammering tools....Why not benchmark to machine shop shot bags and design one for the construction industry...... Reference machine shop shot bag......Any specifics to look for in a leather shot bag for some REALLY basic practicing!? We are restoring a WW II biplane and I want to do the metal work. The Tinman Respondent:...Yes, the bag should made of carefully selected 4 oz.leather, and not Naugahyde or bargain leather. It should be square, as the square ones offer more area for the measure than the round, and the edges can be used for working radiused bends. The round bags tend to leak sooner because the hammering pressure builds only on the one long seam, whereas the square bags allow the corners to absorb the hammering pressure as well. The square bag is easier to hang on to when moving, important if it is large and filled with lead shot. The seams on our design are on only three sides, as the fourth is a fold, and the fold allows really heavy work in that spot without the worry of leaking a seam. Finally, the TM Technologies design is the only one with a positive locking closure at one corner, which both allows re-filling when the bag stretches and loosens with use, and yet without opening and leaking during use, something Velcro will just never handle....

340. Lifting tongs for wood studs and posts

This idea is an ergonomically designed set of tongs, one for left and one for right hand. These tongs fit in the palm of the hand and look like scissors except the ends are specially designed for grasping typical wood studs sizes 2 x 4 to 2 x 10'S and small posts 4 x 4....Working with both tongs the consumer will appreciate the sure grip and easy lift feature of this tool. This tool relieves stress on the body around the hands.......

341. Weatherproof post it notes & pen for construction

This idea looks like the normal post it notes stickers from 3M and pen except my version is waterproof. ..The sticking power of this version to be very high so the consumer can leave it on the job site item and not fall off...The pen is waterproof so the ink will not run...This product benchmarks the 3M "Post it notes" and underwater divers drawing boards....Now in rain or shine the consumer can stick on notes to there construction work and people who are moving can leave notes on there furniture.....

342. Construction push pin with molded finger tip

This idea a heavy duty ergonomically designed push pin to be used on construction project to mark items needing attention... This pin is strong enough to be pushed into wood, drywall, flooring, roofing etc...This pin is very narrow yet extremely durable , the head is molded like finger tip for ergonomics and varies in color per user choice.....Now the consumer has a push pin for construction projects where they can mark off items of construction....The exact dimensions and material of this construction push pin to be determined per experiments.....This product idea benchmarks to office push pins used on cork boards and acupuncture where a very thin and strong need/pin enters the human body.......

343. Residential version of Gunite sprayer and tank

This idea takes the sprayed on concrete Gunite and brings it down one level to homeowner DIY use...This idea is a small sprayer with holding tank for the Gunite powder...Now the typical consumer can beef up certain aspects of there property without having to hire a contractor...The small contractor now has a reasonable tool without having to rent expensive equipment.....Reference..Areas where Gunite is used:.Concrete repair,Slope stabilization,Building stabilization and foundation repair,Culvert and sewer lining,Tank lining,Retainer walls,Swimming pools,Dams...More reference information: Gunite is cost effective, it allows us to place the concrete without forms and at a much faster rate than conventional methods. For example we use gunite to fill the mortar joints between granite walls. What would take masons days or weeks to do we can do in hours. Gunite can be used to place concrete overhead. For example we have used Gunite to repair the underside of slabs in parking garages, paper mills and water treatment plants.....

344. Oven glove material to keep power tool cool

This idea takes the famous oven glove material were people can stick there hands inside a hot oven and still keep them free of burns, but now applies it to tools....This idea takes this material and coats power tools around the housing, on the exterior to keep the users hands comfortable when operating, especially on heavy work loads....I feel the users will appreciate this feature....

345. Cooling towers onboard heavy duty power tools

This idea also uses NANO technology to add small cooling towers on board the heavy duty power tools to cool them down while in a heavy workload mode...These cooling towers would be strategically placed on the tool and more than one tower can be used on the tool...This is not an add on, it would be part of the tools design...This cooling tower works like a cold hair dryer... Ergonomically designed of course to fit the tool design....

346. Car horn onboard heavy duty power tools..

This idea integrates a button and car sounding horn inside the most heavy duty power tools...The use is of course like that of vehicles, to alert somebody of a heads up situation...This is not an add on to the products, this would be integrated inside the products with a NANO module technology...The horn button is on the outside of the tool....

347. Portable robot for general construction work

This is a robot that lives on the back of a big rig truck..The truck carries all the supplies needed to keep this robot going... The truck is taken to a typical location for construction work ...This robot is programmed on site for that certain work, maybe to assemble each room of a house as the carpenters would, except this robot can work 24 hours a day, 7 days a week with exact accuracy but does not drink beer.....This robot is not limited to rough frame, it can extend out to lay tile on roofs, anything on top of ground construction....What a future machine!........ This idea came out of watching the many robots in a typical automobile factory but together a car...Why not take the robot to the site to replace heavy backbreaking work......

348. Rebar tying machine

This idea is to have an American version of a rebar tying machine. The American version can be more ergonomic...We do a lot of concrete work here in southern California and I just do not see this tool...This idea benchamrks to the chinese rebar tying machine from YONKANG HONGTIAN INDUSTRY & TRADE CO. LTD from Province:Zhejiang, China......Product spec..Battery: 9.6V Ni-Cd 2500mAh or Ni-mH 2800mAh Charger: 1 hour quick charger, 110-240V, 50/60Hz Max. Tying Diameter: Dia. 40mm Max. Number of Tying Knots: (one Knot use around 70cm wire) 700Pcs for Ni-Cd battery 900Pcs for Ni-mH ....

349. Oscillating hand held knife W/ Dyson steering ball

This idea is to use NANO technology and make a hand held oscillating knife..I believe the consumer would appreciate the ease in cutting..This knife would also incorporate the Dyson vacum cleaner ball...Now the consumer has a super knife with incredible rolling/maneuvering action................Background info benchmarks to the "Fein" company SuperCut Construction with oscillating movement for universal deployment. Optimally compiled sets for glaziers/window installers, carpenters/interior fitters, tilers/plumbers, and sealant repairers. Fast, reliable, and easy-to-use system. Infinitely variable electronic cutting frequency control. background on Dyson:......The latest Dyson upright vacuums sit on a ball to maneuver around furniture and obstacles....

350. Trash strainer net for job sites

This product idea is to help the DIY and construction people keep there rough job site relatively clean of flying junk...This product is a very lightweight net with two staggered layers of netting on top of each other..Now the job site can breathe and still protect from flying junk and animals....The size of this net is big, enough to cover a twenty by twenty foot room for example.. The key is lightweight and easily hosed off with water....

351. Angle position tool w/roller for glue guns

The idea for this tool is for the consumer to place/sit glue guns in this tool and be able to glide there gun at a certain degree, for example 30 degrees. This tool has a roller extension attached that rolls on a given surface, allowing the consumer to aim there glue a bit better and have a more comfortable feeling doing so....I notice people really mess up with there glue guns, sometimes it's pretty funny, they even get the glue in there hair.....In summary this tool is a small frame with extension arm that has a roller...A simple design .....

352. Electric iron with quick cure tile grout

This is an electric iron ergonomically designed to roll on floors, specifically over the grout between freshly layed tile..The purpose is for this new grout to cure quickly instead of hours.... This idea came from watching tile layers do there work and then waiting many hours for the tile gout to dry...Let's develop a new fast methodology for the tile setters to do a faster job...This idea involves a new grout ingredient to work with heat for fast cure. It also involves a new iron to easy roll on surfaces...

353. Integral power tool gyroscope

This idea is for the consumer to hold a heavy drill in there hand and have it automatically balance itself at a 90 degree against whatever object is being drilled...On a power saw the same thing, the tool would balance itself to cut at optimum position....This idea benchmarks to Ship gyroscopes and nano technology...This advancement would help the consumers in there more precise work and less eye work in positioning power tools...

354. Handles on power tools to center gravity in hand

This idea adds handles to the bodies of your power tools. The reason for this idea is to suggest that besides the hand grips on your heavy tools there exists another area which better balances the tool in the consumers hand...Why not add a formed handle with the frame of the typical power tool...This is not an add on feature, this is a part of the body design...Now I can carry your heavy power tool (which i totally love) around in a better balanced zone with less stress on my hand..This is definitely an ergonomic improvement for your future designs......

355. The three finger glove

This idea is to fashionably design with the correct spandex type material a three finger glove covering the thumb, index, middle finger and half the palm of the hand.....Other finger combinations can also be designed...I notice our workers as well as everybody else for that matter cannot keep gloves on for to long, they get to hot and the hand seems to want to breath, especially on detailed specialized work...This idea advances yet simplifies the concept of gloving the hand and fingers....For example weight lifting gloves have the fingers removed.....

356. The double headed double sided drill

This idea is a typical drill except it has two heads with chucks opposite to each other...On one side the drill chuck can hold a drill for example...On the other side it can hold the tap.......Now the consumer can drill drill then flip the drill over and tap with the other head....Please let me know if you cannot visualize this...I noticed the guys in our shop spend a lot of time drilling, then taking out the drill and reloading with a tap.....Another example is to chuck a .156 diameter drill on one side and a phillips head screwdriver on the other, for the number 10 sheet metal screw to be installed..We do this combo a lot in our shops...The on and off switch can be on top of this product because the handle is constantly being flipped over....I believe this product to be a time saver on assembly lines....

357. Electronic sub systems on high end power tools

This idea is to add back up electronic systems for your high end power tools..The phrase "twin" appeals to customers of high end products...As a young man I got to sell Porsche Sports cars for Vasek Polak in Hermosa Beach California...I remember our customers being delighted when they heard our Porsches had "twin ignition systems"...The concept of having a back up in case the main system fails is perfect for today's society...This connotation is a great feature to add to your products.....

358. Automatic reel out line for hanging flashlights

This product is for the consumer to stand at floor level and be able to automatically extend a line with hook to a higher level object to hang a flashlight. In some situations you want to hang a flashlight to do work....Why not offer a fishing reel type product with automatic extending line that becomes rigid once it leaves the reel, full extension could for about 8 feet...The end of the lightweight metal wire can have a hook on one..The main reel box has velcro straps to attach any flashlight or lantern....For example, I am in a rough frame building and need lighting from above, I simply extend my line to the ceiling rafter and hook to it...I now have a flashlight hanging from above and I can do my work...Memory metal might be a great material for the extending line....If my description does not make sense I can prepare a sketch......

359. Integral air hose hookup for power tool cleaning

This idea adds an air port hookup to power tools so the consumer can hook up an air hose and blow out the dirt and grim.... Air pressure to be regulated of course with this new port...This port becomes an integral part of each tool when it is made at the factory, this is not an add on.....

360. Radio flyer toolbox - Retro design

This product idea designs a toolbox like the old fashioned radio flyer wagon...The retro aspect of this product will bring smiles and feeling of the past for consumers....With this proposal we keep the bottom portion of the wagon with handle and all... The top portion for storing tools to be of fine wood with ergonomic features for modern day tools...The key here is a retro look (Red paint, wheels,handle,pan,wood rails/drawers)......Reference: From Radio Flyer The Radio Flyer Town & Country Wagon is the wooden, large-size wagon with Radio Flyer appeal and wooden durability! Although a wooden wagon, the Radio Flyer Town & Country still includes amazing features like an extra-long handle for easy pulling that tucks under for easy storage, durable steel wheels with rubber tires for a smooth, quiet ride, and no-pinch ball joints to keep kids' fingers safe. In addition, the Town & Country Wagon includes bright red removable wooden sides, as well as added safety measures like handle control of turning radius to prevent tipping and a natural wood body with a smooth, sanded, non-toxic finish. With features like these and a tradition everyone knows and loves, the Radio Flyer Town & Country Wagon is the wooden wagon that will keep up with any active child! Body measures 36"x16.5"x9.5," and wheels measure 10"x1.5." ...

361. Flexible chalk line for contour marking

This idea benchmarks to construction chalk lines except this uses a flexible metal line which is rough contoured by the typical consumer.....This new line has a mechanism attached which has two purposes....First purpose, the consumer (by hand) lays out his rough contour to be marked. Now he runs the mechanism along the contour line, as this mechanism is being pushed by hand it also smoothes out the flexible line which the consumer has contoured...Second purpose is to load a marker on the mechanism and run it again along the approved contour, now the surface is marked with the marker or chalk pencil....Now the consumer has a tool to mark contours....Since the line is flexible it simply rolls back after marking (by hand)....If my description does not make sense I can make a sketch.....

362. High tech fluids application gloves

This product idea is a pair of gloves with integrated fluid release system and finesse finger tips for application of fluids. This product is for the artist craftsman...One glove can be for glue while the other glove could be for solvents...The fluids packet is stored inside the glove at the wrist level which is refillable. There is a fluid line running to the index finger. Upon the touch of a button from another finger the fluids get released at the very tip of the index finger....This would be a high tech way of gluing or cleaning something...It's meant to be for the very detailed craftsman/artist...The material for this glove to be comfortable to the user....The batteries are also stored at the wrist area...The other finger tips have the special tips but only the index finger has the fluids release........

363. Tool for lifting lumber into vertical position

This idea benchmarks to the famous BOWFLEX exercise machine, using those resistance power rods in a different manner....As seen in my sketch the power rods are now used for the consumer to be able to pick up horizontally stacked lumber into an upright position....A special clamp device is used at the end of the bars to clasp the lumber...No power to be used with this tool....The consumer moves the power bars around by hand with an attached handle...The base has enough weight to move lumber. The base also can have wheels and a brake for ease of movement. Now the contractor/framer can rest his back a little bit when raising studs.....

364. Super glue to save knots inside cut lumber

This idea is to find a super glue with unique formula to fuse tree knots in lumber thus saving those pieces of wood. I am talking about reasonable knots, not the worst cases...In today's world of deforestation we must find ways to save as much natural resources as possible....I believe this idea is to help mankind instead of a money maker....Please read the last paragraph in particular....Reference material regarding knots in trees:....Knots materially affect cracking (known in the industry as checking) and warping, ease in working, and cleavability of timber. They are defects which weaken timber and lower its value for structural purposes where strength is an important consideration. The weakening effect is much more serious when timber is subjected to forces perpendicular to the grain and/or tension than where under load along the grain and/or compression. The extent to which knots affect the strength of a beam depends upon their position, size, number, direction of fiber, and condition. A knot on the upper side is compressed, while one on the lower side is subjected to tension. If there is a season check in the knot, as is often the case, it will offer little resistance to this tensile stress. Small knots, however, may be located along the neutral plane of a beam and increase the strength by preventing longitudinal shearing. Knots in a board or plank are least injurious when they extend through it at right angles to its broadest surface. Knots which occur near the ends of a beam do not weaken it. Sound knots which occur in the central portion one-fourth the height of the beam from either edge are not serious defects. Knots do not necessarily influence the stiffness of structural timber. Only defects of the most serious character affect the elastic limit of beams. Stiffness and elastic strength are more dependent upon the quality of the wood fiber than upon defects in the beam. The effect of knots is to reduce the difference between the fiber stress at elastic limit and the modulus of rupture of beams. The breaking strength is very susceptible to defects. Sound knots do not weaken wood when subject to compression parallel to the grain....

365. Special tack cloth to clean hand & power tools

This idea is to sell separate rolls of tack cloth for cleaning the hand and power tools....I feel the consumer would appreciate a special cloth for the expensive tools, even for hammers. A lot of people are neat freaks with there tools.......General background on tack cloth: Tack cloth is a sticky (or tacky) material used for removing dust from a surface prior to finishing it with paint, varnish, or some similar product. Tack cloth is typically used in woodworking, but can be used in other applications as well. Tack cloth works by causing the dust, dirt, and wood particles to stick to the cloth as it is wiped over the surface of the material being cleaned. Tack cloth is sticky enough to pick up the dust, but not so sticky as to leave behind a residue on the material being prepared for finishing. Tack cloths are usually about 12 inches square (25 cm), although the actual size may vary quite a bit. They are typically used after sanding a woodworking project, just before applying the finish....

366. Fire hardening machine for wood studs and posts

Fire hardening is the process of removing moisture from wood by slowly and lightly charring it over a fire. This makes a point, like that of a spear, or an edge, like that of a knife, more durable and dangerous....My idea is to produce a machine in which a typical wood framing stud or small post can be run through in order to fire harden it's edges...The makes for a stronger structure in whatever is being built...The proposed fire hardening machine could look like a portable pass though mill machine...

367. Shape memory alloy for hand held tape measure

This idea brings the new memory metal technique into the typical tape measure...Now the consumer can measure contours as well, keeping that contour intact after measuring the object... The exact uses by the consumer are endless with this product idea...The trick is the methodology used to spring the tape back into a straight position..Reference material:..A shape memory alloy (SMA, also known as a smart alloy, memory metal, or muscle wire) is an alloy that "remembers" its shape... One Way Memory Effect When a shape memory alloy is in its cold state (below As), the metal can be bent or stretched into a variety of new shapes and will hold that shape until it is heated above the transition temperature. Upon heating, the shape changes back to its original shape, regardless of the shape it was when cold. When the metal cools again it will remain in the hot shape, until deformed again. With the one-way effect, cooling from high temperatures does not cause a macroscopic shape change. A deformation is necessary to create the low-temperature shape. On heating, transformation starts at As and is completed at Af (typically 2 to 20 °C or hotter, depending on the alloy or the loading conditions). As is determined by the alloy type and composition. It can be varied between -150 °C and maximum 200 °C. [edit] Two Way Memory Effect The two-way shape memory effect is the effect that the material remembers two different shapes: one at low temperatures, and one at the high temperature shape.A material that shows a shape memory effect during both heating and cooling is called two-way shape memory.This can also be obtained without the application of an external force (intrinsic two-way effect). The reason the material behaves so differently in these situations lies in training. Training implies that a shape memory can "learn" to behave in a certain way. Under normal circumstances, a shape memory alloy "remembers" its high-temperature shape, but upon heating to recover the high-temperature shape, immediately "forgets" the low-temperature shape. However, it can be "trained" to "remember" to leave some reminders of the deformed low-temperature condition in the high-temperature phases. There are several ways of doing this....

368. Benchmark Stereo lithography machines for hardware

This benchmark is to study between Stereo lithography machines and the United States Navy machine shops....It's said that the US Navy can create there own parts from a Stereo lithography...This idea is to have a machine capable of producing any hardware part needed from a programmed menu...I believe a lot of money could be saved by companies in producing there won hardware items...Reference material:........Martello has fifteen years experience of procuring rapid prototyping SLA and SLS parts using a wide range of technologies. We now have our own in house stereo lithography machine (SLA 250) producing 3D model parts in DSM Somos Watershed 11110 resin in 0.1mm layering. We are also happy to provide expert technical advice and source the best quality, process and timescales to suit your needs if we cannot produce the parts in house. ....

369. Portable hardware vending machine for job sites

This idea is a portable vending machine for small hardware items..This vending machine can carry dozens of hardware items needed to finish off construction jobs....Unlike other tool vending machines this one is dedicated for the construction and home improvement people...This design can also be sold to hardware stores for people needing small quantities of hardware...At the hardware store no more waiting in line for an hour to buy a couple of screws, with this machine it's all done in one second.... At the job site it's nice to know all small hardware items are located neatly in one vending machine...

370. Multiple spindle drill head for power drills

This idea benchmarks to the industrial multiple spindle drillers and brings that technology down one notch for the construction people...The exact array of this proposed drill head can vary but with the same feature of being able to attach to the main drill....This tool can also be made to attach to the free standing drill press used by small shops and construction people... The results of this tool is of course time saving..........Please see below for reference material: Multiple spindle heads are designed for use with most drilling machinery. Multiple spindle heads can, almost immediately, double or triple your drilling operations by simultaneously drilling countersinking, reaming or tapping two or more holes in one operation. Multi-spindle heads have proved to be the most versatile method to drill and/or tap close spaced holes. Multiple Spindle Heads Adjustable Multiple Spindle Drilling Heads - Fixed Pattern and Special Purpose Heads Adjustable multi-spindle drilling heads can convert an unproductive single spindle drilling machine to a versatile multi-spindle drilling machine Both adjustable and fixed style multiple spindle heads with the adapters to fit manual, automatic and fixture type drilling equipment are available. These multiple spindle heads are high quality and priced to be affordable. Provide some basic dimensions, as shown on our Application Sheet, for a prompt quotation. Adjustable style multiple spindle heads for drilling and tapping are usually stock items. Adapters are machined in about two weeks. Multiple Spindle heads are sometimes referred to as: cluster drills, tandem drills, or gang drills....

371. Tool to shatter existing metal connections

The idea of this power tool is for the consumer to be able shatter an existing metal connection which is very hard to undue by conventional methods (whether it be a nail, bolt, screw etc)... This tool on one end contains a cartridge of liquid nitrogen which is released via a trigger (in tiny amounts) on the other end a hammer so the consumer can tap the object until it shatters away and loosens the connection......This product idea is not only for the construction field...thank you.....Please review the reference material below: Cold Metal Shattering Metals can become very brittle at low temperatures. However, my guess would be that one would have to throw a wrench very, very hard to break it. My response to an earlier question about why metals become brittle when immersed in liquid nitrogen follows. Actually, many metals can become brittle at temperatures well above that of liquid nitrogen (-196 deg C or -321 deg F). This tendency to be brittle (i.e., fracture under impact) is referred to as a metal's "toughness" and this toughness is temperature sensitive. When a metal is ductile, it can bend and stretch. This change in shape is accompanied by actual translation or flow of the metal at the atomic level. As the temperature decreases, it becomes more difficult to break these bonds and, consequently, easier to develop stresses in the metal that can lead to actual fracture, rather than flow. The crystalline structure of the metal, along with many other factors, influences the temperature at which that metal becomes brittle. Metallurgists have learned to manipulate steel composition to achieve a desired temperature sensitivity. For instance, steel alloys high in nickel are used in cryogenic applications because they are more resistant to becoming brittle at very low temperatures. The effects of temperature on metal toughness can be critical in many applications. For instance, it is hypothesized that the sinking of the Titanic ocean liner might have been averted if the steel in the hull had had greater low-temperature ductility. Tests on hull samples from the Titanic retrieved in recent years indicate that the steel had a high sulfur content, which caused it to become brittle at temperatures as high as -1 deg C, which is substantially above the freezing point of salt water. As a result, when the Titanic struck the iceberg, the steel in its hull fractured rather than deformed, causing the fatal gash. If the metal had just buckled, it is possible the ship would not have sunk. These properties were not well-understood or appreciated until the 1940's. I have not heard that, but I would not be too surprised. Most substances become brittle as the temperature decreases, in addition, uneven contraction sets up stresses within the object. It is entirely conceivable that a metal tool could become sufficiently brittle at -40 C. ~ -40 F. to break apart if dropped...

372. Tool for projecting 13 inches with laser readout

This idea uses a laser beam with is incrementally cut into for example 13 inches. The trick here is the readout on whatever surface the laser hits, displaying the numbers....For example, as shown in my sketch, if I am standing on the floor and shoot my laser to the ceiling rafter to see how big it is...Once the laser hits the rafter I adjust my laser start point to zero and see that the rafter is 6 inches high...The uses for this tool are many and to cut down time when needing quick measurements....This tool is hand held like a tape measure, actual dimensions to be determined. I see too many people wasting time needing small measurements on odd positions. Why not offer laser technology with readouts for quick visual measurements check......

373. Compressed air backpack

This idea benchmarks to the paintball game compressed air tank which is worn on the body...I am thinking of a ten pound tank which is worn on the back as shown in the sketch...The tank to be as light as possible so aluminum could be used for the housing...A few ports are available on this design for other users to connect to....This idea is for light jobs needing quick solutions with air tools........Please read the following reference material for the paintball air tank.."Product Description The 72 is one of Pure Energy's pro level air systems. This system combines a popular bottle and the top-of-the-line Pure Energy Reactor regulator. The Pure Energy Reactor regulator has a preset output pressure set between 800-900 PSI. This air-system works great with any marker on the market, from Piranhas to Angels. Because Pure Energy Reactor regulators are now re-buildable, this system is one of the most user-friendly, screw-in air systems in paintball. On most markers you can expect to get 700-1100 shots from a full fill. The exemption for Pure Energy's carbon fiber bottle means you only have to have it re-hydro tested every five years. With the CFT piston and a streamlined setup, this air system is both extremely consistent over the chronograph and has very little drop off as you shoot the gas out of the tank (life of the fill). These are attributes sometimes found in much more expensive air-systems packaged into a smaller, easier-to-use, screw-in system. Feature Detail: * 72 cubic inches lightweight aluminum composite * Consistent flow technology piston * Mil-spec Belleville springs * Dual safety rupture plugs for both bottle and downstream safety * Mechanical gauge * Nickel plated brass bonnet * Certified fill nipple and dual locking set screws Method of Operation: Preset screw-in air system Bottle: 72 cubic inches lightweight aluminum composite Power Supply: Compressed air or nitrogen Length: 11 inches Weight: 3.1 pounds...

374. Self lubricating cutting tools for all power tools

This idea benchmarks to self lubricates bushings ...Please read the reference material below...The main idea is to have high speed cutting tools last longer and cut better via lubricating, in fact lubricating themselves...The other angle to this idea is to embed temperature sensors to power cutting tools and via a an internal NANO mechanism have the tool shoot a burst of lubricant to the cutter (wheels, blades, drills etc.).....................Below is the reference material: hese bushings are called 650 Series and are designed from either cast iron or bronze-based metal to withstand high load. Finely finished radial holes are embedded with a solid lubricant covering up to 30% of the working surface. During operation, this lubricant automatically deposits a film onto the rotating surface of the shaft. This makes the bushings particularly suitable for applications requiring extreme temperatures, high pressure or corrosive chemical or water conditions where no lubrication can be introduced. AST's new self-lubricating bushings offer superior performance without lubrication under severe temperatures (-100?C to +300?C for bronze metal) and (-40?C to +400?C for iron metal) and can handle a maximum load of 150N/mm2 and maximum speed of <1.00 m/s. AST's 650 Series of Bushings is suitable for such high-load, low-speed, high-temperature applications such as: iron and steel manufacturing equipment, automotive machinery, injection molding machines, food processing, paper, chemical and textile machinery and other heavy industrial applications. A broad range of sizes and shapes are available, all manufactured in certified ISO9002 factories. In addition, the 650 and 250 Series Bushings offer: § Self lubrication capacity and extremely long life § High load, anti-wear and low-friction attributes § Suitable for extreme temperatures and low speed § Chemical resistant and anti-corrosive § Suitable for reciprocating, oscillating or intermittent motion where it is difficult to form an oil film The 650 and 250 Series Bushings are available in the USA exclusively from AST Bearings. For more information and free literature about AST's 650 and 250 Series Bushings, contact: AST Bearings, 115 Main Road, Montville, NJ 07045, or call: (800) 526-1250 NJ, (800) 227-8786 CA ....

375. Semi circle treads for ladders

This idea is to design a step ladder, no matter what height, with stair treads designed in a semi circle (half round)...Reason for a semi circle tread is for better positioning around a 180 degree field of impact by the foot....This idea adds a little bit more ergonomic touch to the typical ladder tread....If you really think about it it's a little uncomfortable when climbing a ladder and working on one, lets make it feel better....

376. E-Z glide hand truck for door hanging

This product is an ergonomically designed hand truck that actually glides on ground surfaces...This product is designed to move doors around and help when the door is being hung...The age old task of hanging doors can be eased with this product...A cylinder covered with the special low coefficient of friction material can glide on any surface. The lip which carries the typical door is hooked down so it will be easy to hang the door on it's frame.. Reference Data: "EZ Moves Furniture Slides Can Help Workers Perform Their Jobs Safer and Easier After the success of its EZ Moves Furniture Slides, Simtec Company is giving a real "push" to its professional product line designed to help workers in the janitorial and hospitality fields reduce back injuries. Candy Holsing, Director of Marketing for EZ Moves, who has a background in occupational medicine, said it's her personal mission to eliminate back problems for workers related to heavy lifting and moving. "When you have a repetitive or heavy lifting task every day, such as moving furniture to clean, it can take a toll physically," said Holsing. "Often these workers cannot afford the high costs of healthcare or to be out of work for long recovery periods." The slides, in both EZ Moves Pro models and the 4 ft. long ski-like slides, are designed to reduce friction on carpeting and most hard surface floors"........

377. Tool to dislodge deeply embedded nails

This tool bores a hole around the head of the typical embedded nail which is hard to remove...Flip this tool around and use the hammer claw to dig into the bored hole perimeter to engage the nail head and pull out in a conventional hammer claw manner. I believe this is a great tool to help with the boring job of removing deeply embedded nails....

378. Nail positioning tool before & after hammering

This product is a simple hand held tool that the consumer uses to position a typical nail before hammering down then slips out after afterwards....The nail is gripped via a spring molded inside the plastic slit area, acting like jaws to clasp the typical nail....I believe this tool is great for the folks who's aim is not all that great...This tool also gives pinpoint accuracy for nail positioning on a surface...The user would hold this tool with the left hand and then nail down with hammer on the right hand...

379. Workman's overalls with spring padding

This idea fashionably designs workman's overalls with a thick padding and springs inside to cushion the consumers body when working in odd positions...The actual overall can vary in design to cover more areas than others....Now the consumer can work in odd positions and not get so banged up or fatigued as before....

380. The backpack toolbox

This idea designs an ergonomic modern backpack to hold hand tools...Let's face it, a lot of people love to carry pack packs everywhere....Why not design a very cool backpack to hold hand tools and small hardware items....I'm positive the handyman will love this idea....Benchmark to electricians bags....

381. The Super Vacuum Hose

This idea calls out for a super long vacuum hose to give the consumer the ability to keep a powerful vacuum stationary in one place and run it throughout the construction site or home.....
This idea is mainly for the rough frame mode of a typical construction project...This product might also work its way to the home improvement people and possibly the homeowners if cheap enough....This product installs light relay suction motors in a

novel way inside the entire length of the hose.....These relays act to take the dirt and dust back to the main vacuum station....I can see the neat and tidy people loving this unit......

382. Tool drop guard squares

This product idea are small squares made of heavy duty cushioning material which help hand tools when they are dropped... This product can be simple peel off squares which the consumer attaches to any tool which is dropped frequently or just to safe guard expensive ones....The exact width and height of these squares can be determined after a quick study of tools, especially power tools.....If we can save one tool out there then this product is worth it....

383. Sound deadening glove for construction /Home Improvement

This product idea is an over sized baseball type glove except made from sound deadening material.....The user wears this glove on the opposite hand which is doing work with a tool... For example if I am hammering with my right hand I wear this glove on my left hand and place it relatively close to the impact of hammer and object....The reason for this product is to help anyone doing projects at odd hours or just helping to keep a quiet area.....The other day I was wearing my baseball glove on my left hand but had to nail something down, i left he glove on and noticed it help deaden the sound of the impact...

384. The Nail Sorter

This idea benchmarks to the famous "COINSTAR" coin counting machine except my idea is for nails used n the carpentry business ...This "Nail Sorter" is a small ergonomically designed box with a receiving chute on top...The consumer dumps there jars of nails into the top chute, turns the machine on and the nails get sorted into compartments. Now the user retrieves the grouped nails instead of throwing the unsorted nails away (human psychology)....This product can be multi tiered for more categories of nails to be sorted..

385. Hand held drone surveillance plane for consumers

This idea is for security purposes. Instead of having fixed video cameras all over a typical property why not replace those cameras with one special hand help drone plane with video surveilance..NANO technology comes into play here to take advantage of battery/ fuel savings....The consumer would program a flight pattern to cover there property then this drone plane could circle that property in flight for as many hours as needed....There is nothing better than a birds eye view of something happening on the ground...If the Army can do it why not for consumers too....

386. Embedded NANO material into all hardware items

This idea involves embedding NANO material/technology into every single piece of hardware made at the moment that hardware part is actually being produced in the factories..... Afterwards the typical contractor can scan the job site and get a reading from a portable device of every hardware item actually used on that job....I'm thinking of a laser scanner to give a reading for this certain NANO material/tech embedded inside each piece of hardware, I mean even nails....This idea actually came to me years ago when I found out that American special forces soldiers have embedded chips in there bodies to track there movements...The results of this kind of product/idea would revolutionize the construction industry....

387. Millennium paint can

This product idea for a newly designed paint can with the following features: 1. Square instead of round design...2...Clear see thru walls to check paint levels...3...Special brush dip lid on top with paint swirling feature...4..Rubber feet for height...5.. Fold out paint tray at bottom...6. Integrated paint brush holder to leave the brushes in solvents...7. cushion grip bar for lifting....8. mini water spout at bottom of one wall to release paint onto the fold out paint tray feature...I just think it's time to modernize the typical paint can...

388. Integrated pull out plugs on power strip

This idea takes a power strip for electrical connections that plugs into the typical wall socket and adds pull out reels to each of the integrated plugs in the power strip....Now the user can go over to the power strip on the wall and pull out one or all of the plugs...The distance of cord pull out is determined by the size of this new power strip...Normally the user brings and extension cord to the power strip to connect an electrical item...see attached existing power strip as reference....

389. Portable laser beam wall

The use of this product is to have a visual laser beam wall... This product looks like the cones they use on the freeways to designate work areas except this idea is a tube and about four feet tall with laser beam technology inside....Now the user can fence off areas that need attention or caution or otherwise with lasers instead of other cumbersome barriers. I can see many uses for a laser beam wall.....This product idea is a hollow tube about four feet high with base for vertical stationing. Inside the hollow tube is a laser beam that can shoot out many tiny laser beams in a vertical position through holes in the hollow tube wall maybe half inch on center (done with mirrors from the main laser inside tube).....The other tube with base is used to stop the lasers from continuing on forever....

390. Work gloves with imbedded L.E.D. lights

This idea is to imbed some L.E.D. lights into some work gloves and give it a try...It would be unique to have lights with tiny battery imbedded into heavy duty work gloves...The lights are of course strategically located as not to be crushed when something is being carried in the palm....Reason for the lights is of course to light up areas without having to carry a flashlight...I'm thinking the light s would be imbedded into this new gloves wrist area, top and bottom....I'm visioning somebody carrying a large box and still be able to light a path......

391. Magnetic pickup tool with flashlight at handle

This idea benchmarks to the Sears telescoping magnetic pickup tool which can pick up objects to 5 pounds and has a flashlight at the very tip of the magnet...My idea simply changes the position of the flashlight to the handle, pointing down at the magnet area to achieve a 180 degree light up field of vision... The Sears tool only lights the very tip of the tool with very little field of vision light up area after object is picked up....With my idea we see what's happening after the magnet has picked up the object.....

392. Glow in the dark markings on tools

This idea simply paints glow in the dark marking on tools in unique areas...Whether is be the name or the "ON/OFF" mark, the up and down mark, the teeth on a typical saw, any danger of cutting area etc.....I feel that the glow in the dark markings adds a touch of thought. Situations could arise when the tool marking offers a superb and cheap safety feature, for example when the evening hits the typical job site and the lighting is dim...

393....3 in 1 Safety bullhorn for home emergencies

This product design is a modern ergonomically designed bull-horn which includes flashlights and smoke detector. This product attaches to any wall outlet for recharging it's batteries...When an emergency occurs the consumer has a chance to light the way as well as call out instructions or help via the loud bullhorn. Your neighbors will love you or hate you....This idea came out of watching commercials with manufacturers coming out with simple products again, like the new skill power scissor....

394. Light emitting diodes along tape measure reels

This idea places tiny L.E.D's along the full length of the typi-cal tape measure, running down the middle of the reel to light it up in order for the consumer to have extra help in reading measurements....A button on the tape cartridge can be the on and off button for the LED's....In conclusion what you have is a system to light up any measurement tape reel ...From the ridiculous to the sublime but one never knows what will be in fashion...I notice all new gadgets in the hardware sections sell out......

395. The power tool trigger lock

This idea simply introduces an ergonomic one design fits all lock for all power tool triggers....This idea and design benchmarks to firearm trigger locks as noted below, example of firearm lock:"In this day of lawyers and liability you need to make sure that your firearms are not available to children and unauthorized adults. Of course, a safe is your best alternative but sometimes not practical or affordable. The trigger lock is your next best option. Ours features a three digit combination lock that allows quick access and no keys to ever lose. Over 1000 combinations are available and multiple trigger locks can be set to the same combination for your convenience. Works on almost all handguns, rifles and shotguns"".

396. "V" shaped tape measure reel

This idea takes the extending reel of the typical tape measure and makes it into an upside down "V" shape instead of the straight line.....The new V design is molded metal yet when it winds back up into the cartridge it goes flat, via rollers inside the cartridge, when extended again it goes back into the its upside down V shape ....The V coming out is upside down so the user has two side on which to view there measurements...Now the consumer has a better visual from two side of there measurements since the walls are tilted upwards...This new V shape also gives me better handling control of my tape measure reel with my thumb and index finger... I can make a sketch of this idea if this not clear what I am saying......

397. Incremental holes template on tape measure reel

This idea takes the typical tape measure and adds tiny hole cutouts incrementally along the center line of the tape reel.... The hole cutouts are small, enough for a pencil or pen point to pass through and mark the other surface being measured.... The increments of cutouts can be as desired , for sure every one inch....Now the consumer does not need to pinpoint with there pencil the one inch mark at the edge of the tape measure, they simply put there marking instrument in the center hole of the one inch mark...I notice people spending time on visualizing there increments on tape measures edges, why not make it fool proof...I understand the tolerances are big in construction but still it's nice to know precision is important....In overview of this idea, the tape reel that comes out of the typical tape measure has tiny holes at whatever increment is desired. The user simply marks off measurements using this holes template.....

398. Stud sensor with vertical line laser

Today I was examining stud sensors and thought it would be a great idea to add a vertical laser line to it...When the consumer finds the stud in the wall it would be nice to have a laser line from floor to ceiling showing on that wall...This laser just adds another useful function for the consumer to mark the wall.....A few of the customers i talked to at the store agreed with me....

399. Hurriquake nail positioning tool

This product idea is to help the end user when nailing the incredible hurriquake nails at the two inches on center as specified for max performance.....This product idea is a narrow flat bar about 24 inches long (shorter for home use) with a handle... This product has holes at two inches on center running along the flat bar length of 26 inches...These holes have special rubber boots that hold the nails in place...The idea is for the consumer to load up this product with 12 of your hurriquake nails then place this product over the area to be nailed via the handle... Now the user simply hammers down the hurriquake nails...The special rubber boots go down with the nail then spring up as the nail are driven home....The home version can be smaller, say 12 inches long to hold six nails....To clarify that the user can hand load as many nails as they wish on this product from one to whatever is designed. I might want to load up 4 nails for some small project......The product looks like a typical file but made of different material with holes running down the length with rubber boots attached.....I find it much easier to use positioning tools so i now i get max performance out of hardware......

400. Over the door hanging toolbox

This idea is a more slender toolbox with hooks for hanging over any door...The actual design can vary greatly per preference ..The main idea is to have a toolbox any consumer can hang onto a typical door.......The add can go like this:.."""Utilizing usually wasted back-of-the-door space, this toolbox uses sturdy brackets ... No more tools lying around the floor and drawers! With this over the door toolbox""", ...Gosh with this idea no more wasted drawers full of tools, people can now have clean garages too....

401. Hand truck with adjustable rear angles support

This product idea comes from watching people for many years kinda struggle when lifting and moving heavy objects with hand trucks....why not add a separate set of wheels and adjustable scissor like arms to the rear of the hand truck which the consumer can adjust for whatever angle they desire?...Now they can slide the truck arm under the object as usual then set an angle for backwards tilting and movement forward....This new mechanism acts like a backwards brace so the consumer does not actually have to carry any load...the new mechanism is flat against the truck when not in use so storage is relatively flat...

402. The double barrel multifunction flashlight...

In our modern age of super duty products I thought it worthwhile to suggest a flashlight with two side by side barrels that are a few inches apart. The handle is in the middle, picture an eating fork in your mind for the general design.....The reason for two separate flashlights in for back up if one light goes out, kind of like the dual ignition system on the Porsche automobile concept...The handle for this product is hollow with the cap that unscrews. The hollow handle can have fireproof matches inside... This product has a 24/7 light emitting diode light that never goes out as long as the batteries are good...The double barrels are slightly tilted to the inside by a few degrees to give a better field of lighting. This product

comes with a whistle designed right into the handle cap .... This is also waterproof...The overall feel would be a very durable flashlight for the consumer.......

403. PEZ Dispenser for small hardware items

This product idea is a dispenser like the famous PEZ candy dispenser except our design is for any small hardware item that can be stacked....The purpose of this idea is to conserve small hardware items and have them neatly and conveniently ready for use...Our design could be see through plastic so the consumer can visually see how many items they have left....This basic working of this invention would be similar to the PEZ candy dispenser with the typical consumer lifting a lid to reveal the item being stored, except in a more industrial designed version the the PEZ ....I would like to use the word "PEZ" because it draws memory of the famous candy and a positive feeling......The exact size of this product idea can vary according to hardware being dispensed, example: nuts, washers, maybe even nails....No more messes and a cool way to store your hardware...

404. The modern home escape ladder

This product idea is a very lightweight tube about six feet long and a few inches in diameter. It has two handles in the middle for carrying...It has lightweight bars attached horizontally to the main tube, spaced every twelve inches or so, these will be used as steps...It has two lights, one at each end of this vertical the tube....It has a unique claw design at each end of the tube and attached horizontally, these are used as a sill claw or base for exiting the window escape....It also has a battery pack and two electrical outlet prongs to keep the lights battery charged.....This product with a brace hangs on a typical dwelling wall and connects to the typical wall outlet....When an emergency arises the user determines the gravity of the situation, once this product is taken off the wall then the two lights at each end automatically turn on...Now the user determines if to break windows and escape or use this product as some sort of tool to minimize danger. For example there is a fire in the house and the user is in the bedroom with this tool, they can either escape through a window or use this tool to help other members of the family escape to... The main idea is lightweight so a teenager can use it to...

405. Spray to fight gasoline spills at fill up

This product is a spray that fights gasoline spills on vehicle paints and plastic parts. Used when the typical consumer goes to fill up at the gas station...I notice most if not everybody in the world eventually spills gasoline from the nozzle around there vehicles gas tank funnel.......A typical consumer goes to fill up and as usual spills some gasoline around the gas tank funnel, now they take out this product and spray it at the spill, knowing that gasoline damage on the vehicle has been neutralized....The exact formula of this gasoline neutralizer to be determined...

406. Pull out electrical wall outlet with cord and reel

This idea is a typical wall outlet for a home or business except this product attaches an extension cord (example 25 feet) which is buried with a reel device inside the wall and cannot be seen.... The typical user only views a typical wall outlet but when in need of an electrical extension cord they simply walk over to this special wall outlet and detaches it via a small handle on the outlet and takes it to the needed situation for an electrical connection (as the attached extension cord gets pulled out automatically) ....Once done the user can press a button and the reel begins to wind up the cord...This product would be a kit of course that can fit between the spacing of typical wall studs, say 16 inches wide....Idea came to me when noticing people scrambling for extension cords...Why not just bury one in a wall?...

407. The NEW finger thimble

This product idea is a glove that fits over any of the human fingers...The purpose is for the user to still have great dexterity with the use of there hands while still protecting there fingers from dirt and sharp corners...In other words this product is a glove for the fingers only......The material could be a heavy duty neoprene with tiny grippers molded in for gripping power but still flexible enough to easily put over any finger on the human hand....I just found that complete gloves take away from the movement of my hand...The end user can also put on as many "FINGER THIMBLES" as they desire for the work specified.... With me I notice i use three fingers most of the time and so I would have three of these...My mom is 70 years old and she would like just one of these to be able to open up those hard to open rolled up plastic bags at the grocery stores for fruits and vegetables.....

408. The five in one fire extinguisher

This product idea is a fire extinguisher except this modern version has a carbon monoxide detector, a smoke detector, a heavy duty light and a battering ram at the bottom...It is electronically operated via the typical wall outlet..a special brace to hold this new extinguisher against a wall is attached to the studs of a structure, this brace has a cutout in the middle to fit over the typical wall outlet where the new fire extinguisher prongs connect...This new extinguisher has two handles welded to the main tank, purpose being it can be used as a small battering ram in emergencies........In operation when this new extinguisher detects an alarm it's light turns on, the buzzer from the smoke or monoxide detector sound. Now the consumer decides what type of emergency it is and what action to take, up to the point of using the ram end of this extinguisher to break windows or doors down to help victims, and of course to put out a fire.....

409. Dust & Dirt shield for leaf and bow rake

This is a simple product idea, a shield over a typical leaf or bow rake to prevent dirt and other particles from getting onto the user...This shield might be made of see thru acrylic enabling the user to still see what they are raking....This product is an attachment that fits onto any standard rake, it's sold separately..... This shield press fits onto the very end of the handle near the actual rake....I notice a lot of people also raking dirt when they do there yard, causing dust clouds, why not minimize those dust/dirt clouds with a Shield?.....

410. The "GREEN" tools

This concept are made from biodegrable natural resources,,.....Making biodegrable products helps with our overall environment.....When reviewing all the tool makers in our typical stores I noticed no big corporate giant has actually marketed the GREEN technologies....If the United States Army can do it then why not our great tool makers?.....My idea is to review all materials and procedures for making all products..Then for to make an effort to use as much biodegrable produts as possible......This is a typical example:..."Heat Resistance PLA utensils made from heat resistance PLA in Non-GMO corn starch, this cutlery will withstand temperatures of up to 55 degrees (F). 100% biodegradable and compostable. Lightweight and strength.....

411. Magnetic sorting tray for nuts, bolts, nails etc.

This product idea a simple hand held tray with interior magnetized walls....The purpose of this product is for the typical consumer to dump there jar of nails, bolts, nuts etc. into this tray then hand sort the various categories of items. Since all the interior walls of this product are magnetized then the sorted items remain that way for ease of organization to the user....This proposed tray is also stackable so the consumer can buy a few of them for exact precise organization of such a time consuming issue as that of organizing tiny items...This product idea came to me while watching handymen and there jars of nails,nuts,bolts etc..They simply dump there jars into a towel or at the job site then try to locate whatever small item is needed...

412. Daily electricity usage reader for dwellings and businesses

This proposed product electronically reads the power companies standard electricity usage meter which normally sits outside of dwellings and are hard to read by the typical consumer. This proposed product sends a more detailed usage rate to the proposed device inside the swelling.....Now the typical consumer has a daily usage consumption rate of electricity if they wish... The readout on the proposed device can be in a sine curve like they use at hospitals for ekg heart rate monitors....This proposed device/product can hang on a wall or sit on a counter for ease of reading....something that looks like a thermostat.....This idea is a revised write up of my previous idea with the same name..... Idea came from noticing consumers are sometimes very surprised by there costly (high) electricity bills...

413. Conventional hammer with new arm port

This idea adds a threaded hole to any typical hammer, a port as I call it, the purpose is for the end user to be able to screw on a small arm to the threaded port..now the user has more grip to hit the nail and more stability too, on jobs needing very close quarter nailing....

414. Attachable Laser Pointer for lawn garden tools

This idea is an attachable laser pointer used to define straight lines on the typical lawn when the end user is using a power lawn tool, specifically for hand held power edgers....This new laser would be visible in the daytime and safe for human use... This new product idea can attach to other lawn care products as well...My idea came from watching lawn care people miss hit concrete edges with hand held edgers...Now they can watch the laser line in front of them as they edge the lawns...A more clean and precise method....

415. The tool locator buzzer

This idea adds an actual tiny heavy duty buzzer attachment for use on any tool, this buzzer sounds when a button is pressed, similar to the key chain alarm for locking and unlocking your car......The reason for this product is to actually find tools on the job site...Sometimes we just lose our tools....

416. Hammer head with integral nail straightener – reverse claw

This idea adds a tiny hook like design atop the hammer tip which contacts the nail...This new nail straightener is used when a nail goes crooked on the user and the nail is in need of straightening so the nail may be driven inside the object being nailed..... this new design looks like a tiny shovel with more curvature, does not interfere with the hammering of object but assists the user in not needing to get another tool in order to straighten the nail....This problem of crooked nails and straightening them out for further nailing is an age old problem....

417. Nail holder on hammer handle

This idea is a integrated nail holder right into the hammer handle itself. The nails hide inside but can easily be flipped out with the flicker of ones finger....It's nice to have a few nails available on the hammer....The user can load up the nail holder then go nail up whatever item...

418. Hammer head with integral base design

This idea modifies the typical hammer head forging, adding a flat side to the area between the head and claw...The idea is for the typical user to be able to use the hammer head side to sit the hammer up.....I notice people trying to stand up the hammers by the head side. Why not a bit more material and flat so they can indeed stand the hammer up on a flat surface with the handle sticking up...I know I would like this feature, plus it saves me from reaching all the way down to the floor for my hammer, a few more inches in the air makes a difference when we work all day long...

419. The hydrogen powered lawn mower

This idea adds hydrogen technology to lawn mowers...The savings in fossil fuels can make a small dent in today's global warming problem...

420. Lock jaws clamps for tape measure

This idea adds an interesting locking jaw clamp device to the exposed end of your typical tape measure.....This idea is to help a single user when trying to measure distances over the span of open arm lengths....The end of this new tape measure locks onto whatever feature is the start point, then the user simply walks over to the other end without needing to worry about the loose end falling down... This new locking jaw clamp can also be a product that attaches to all existing tape measures in the market....

421. The personal survival cube

This product idea is like an over sized safe...It can fit one person and protect that person from bullets or fire or water for a couple of hours....The main idea is when a sudden catastrophe happens....The uses can be at work , schools, home....In this world its nice to know there is a place to be safe when a sudden disaster is happening......

422. The 911 porch light

This idea installs a typical porch light on the outside of a typical dwelling...when an emergency arises the user inside the dwelling pulls the porch light from inside the dwelling toggle in a left to right action instead of normal up and down, this action now makes this new porch light into a red flashing and if possible rotating red light...This red light distinguishes this dwelling for neighbors and fire- police to help find the distressed person or to signal "danger at this dwelling"...

423. The talking calculator

This idea simply adds voice recognition to a typical calculator to help the user who has there hands full and cannot perform the typing action on the calculator. also great for engineering people who are designing on computer aided design stations and need to perform high end calculations at the same time...A typical user can get close to this proposed calculator and for example say: " 32 times 34 divided by 10"....The calculator would answer with a voice and display it digitally for the user....the use can continue or wait and look at the digital results...

424. Automatic square foot and volume measurer

This device is used in a typical enclosed space, whether home or business, to measure the length by width by height of a typical enclosed space. All measurements at the same time using your laser beam technology...The user does not need to measure or shoot individual laser beams ...The user simply sets up this device in the middle of the enclosed space, then presses the "measurements" button on the proposed device...Lasers shoot from all four sides plus the top to give the user a reading of length and width and height......This product would be extremely useful for all enclosed spaces to estimate all sorts of work to be done, example: flooring, painting, and general estimating purposes.....Of course the area in front of the beams have to be cleared by the user to the typical end wall to be measured....also measures cubic feet of typical enclosed spaced for Heating and venting purposes...

THE END

Thank you very much for reading my book and I hope it has instilled creative ideas of your own.

If you wish to use any of my listed ideas exactly as written and want sketches and or more details please contact me at GREENS31@HOTMAIL.COM Please use "BOOK COMMENT" in subject line.

Respectfully,

Dino von Noy
Los Angeles California